Adaptations to Terrestrial Environments

Adaptations to Terrestrial Environments

Edited by

N. S. Margaris
M. Arianoutsou-Faraggitaki

University of Thessaloniki
Thessaloniki, Greece

and

R. J. Reiter

University of Texas
Health Science Center
San Antonio, Texas

PLENUM PRESS • NEW YORK AND LONDON

Library of Congress Cataloging in Publication Data

International Symposium on Adaptations to Terrestrial Environment (1982:
Chalkidiké, Greece)
Adaptations to terrestrial environments.

Includes bibliographical references and indexes.
1. Adaptation (Biology)—Congresses. I. Margaris, N. S. II. Arianoutsou-Farag-
gitaki, M. III. Reiter, R. J. IV. Title.
QH546.I57 1982   574.5′222   83-13713
ISBN-13: 978-1-4615-8347-9   e-ISBN-13: 978-1-4615-8345-5
DOI:10.1007/978-1-4615-8345-5

Proceedings of the International Symposium on Adaptations to Terrestrial Environment,
held September 26–October 2, 1982, in Khalkidiki, Greece

©1983 Plenum Press, New York
A Division of Plenum Publishing Corporation
233 Spring Street, New York, N.Y. 10013

Softcover reprint of the hardcover 1st edition 1983

PREFACE

The present volume contains selected papers of the International Symposium on Adaptations to Terrestrial Environment, held in Halki-diki, Greece from Sept 26th to Oct 2nd, 1982.

The meeting was designed to consider the means as well as the mechanisms whereby organisms adapt to their environment. The papers presented dealt with a large variety of species from insects up to and including mammals. What became apparent during the course of the meeting was the incredible variety of means that organisms use to survive in their particular environmental niche. The ploys utilized are almost as numerous as the number of species investi-gated. This will become clearly apparent in the accompanying manu-scripts which are published in this book. The Editors allowed the authors of the accepted papers great leeway in terms of the thorough-ness of their contributions. Some of the presentations contain exclusively new findings, whereas others extensively review the existing literature.

The Volume is divided into two parts: Invertebrates and Verte-brates. The first provides information on adaptations of inverte-brates on environmental stresses (such as low or high temperatures and water deficits) from the physiological and/or biochemical points of view as well as behavioral responses resulting from their life strategies and interactions with other organisms. In the second part papers selected deal with vertebrates. Adaptations to special environmental factors such as light and temperature are discussed as well as behavioral, physiological and biochemical solutions to problems imposed.

The Editors would like to express their thankfulnes to UNESCO for sponsoring the Symposium in the frame of Man and Biosphere (MAB) Program; to all scientists who have contributed papers in this volume; and to Mrs A. Karamanli-Vlahopoulou for her patient and skillful typing of the manuscript.

N. S. Margaris

M. Arianoutsou-Faraggitaki   R. J. Reiter

Division of Ecology   Department of Anatomy
Department of Biology   The University of Texas
University of Thessaloniki   Health Science Center at
GREECE   San Antonio, Texas
                                              U.S.A.

CONTENTS

PART 1

Invertebrates

LIMITING SIMILARITY IN ROVE BEETLES (COL. STAPHYLINIDAE) OF A HABITAT INLAND

W. Topp

Lehrstuhl für Tierökologie der Universität Bayreuth
Universitätstr. 30
D-8580 Bayreuth, West Germany

INTRODUCTION

In Central Europe most individuals of soil and litter dwelling beetles belong to rove beetles (Staphylinidae). Communities of rove beetles usually include many species co-occuring within small areas. Do these species show similar ecological requirements, are species independently or randomly distributed or is there any character divergence which can be interpreted as an evolutionary response to reduce competition? Interpretation of community structures in vertebrates have led to different results (Connor and Simberloff, 1979; Simberloff and Boeckler, 1981; Gilpin and Diamond, 1982; Bowers and Brown, 1982) and also results of investigations on arthropod species communities are contradictory. The aim of this study was to examine resource utilization of rove beetles and to test whether species niches are overdispersed. Niche overdispersion might occur when a species minimizes its interaction with all others. The null hypothesis was that niches are distributed at random.

METHODS

Data were collected in a small inland dune ecosystem of ca. 100x200 m in size surrounded by bogs and marshland. Pitfall traps and extraction funnels were used. Samples were taken in 1970-1976 using a grid in each vegetational type (Topp, 1979a).

Horizontal distribution was measured as niche overlap according to formulae proposed by Mac Arthur and Levins (1967) used in the symmetric form (Pianka, 1974). To develop a dendrogram an unweighted average linkage cluster algorithm was applied (Sneath and

3

Sokal, 1973). These methods were only used for a quantitative measurement in comparing sample sites.

Most species were reared in the laboratory under different regimes (Topp, 1979b). Information on breeding seasons and annual time segregation was obtained by rearings, by dissecting females from litter samples and examining them for presence or absence of developing and mature eggs, and by collecting larvae from the field. Information on food type partitioning was obtained by observations under field and laboratory conditions.

Seasonal activity of the two species *Zyras humeralis* and *Drusilla canaliculata* (Fig. 5) was determined by Cornell Ecology Program 12 (Gauch and Chase, 1974). The plotted curves are standardized on the modus=100%. *Z. humeralis*: allopatric=628 ind., variance accounted for 99.6%; sympatric=145 ind., var. accounted for 95.7%. *D. canaliculata*: allopatric=176 ind., var. accounted for 79.4%; sympatric= 99 ind., var. acounted for 92.1%.

RESULTS

A total of 4581 staphylinids, representing 128 species was caught. As in most assemblages, the majority of species occurred in very low numbers. When dominance values are used to construct a species abundance curve, it appears that the data best fit a log-normal series. The resulting curve (Fig. 1), which includes almost the same number of specimens for several species, flattens out which suggests that several more or less independent processes determine species abundance (May, 1976).

Only the most common species of staphylinids were considered here, i.e., 33 species each with more than 10 specimens. The same species were not found within all habitats of the dune system and obvious differences occurred between adjacent sites. *Mycetoporus brunneus* was a typical species of the spruce stand (habitat assemblage (=h. ass.) No 1 in Fig. 2), *Oxypoda togata* was dominant in the driest areas which were characterized by the Cyperaceae *Carex arenaria* and the moss *Rhacomitrium canescens* (h. ass. No 2), *Quedius aridulus* only occurred in sites which were covered by *Calluna vulgaris* (h. ass. No 3), *Platydracus fulvipes* was restricted to the wettest areas bordering the surrounding swamps (h. ass. No. 4), and *Conosoma marshami* was found only in the litter of the pine stand (h. ass. No 5). Some species, e.g., *Othius punctulatus*, spread over a larger area but only occurred in the shaded zone of the trees and avoided the open sites.

The dendrogram (Fig. 2), using average linkage clustering (Sneath and Sokal, 1973), illustrates how species in the same dune differ in horizontal distribution and niche overlap, a pattern

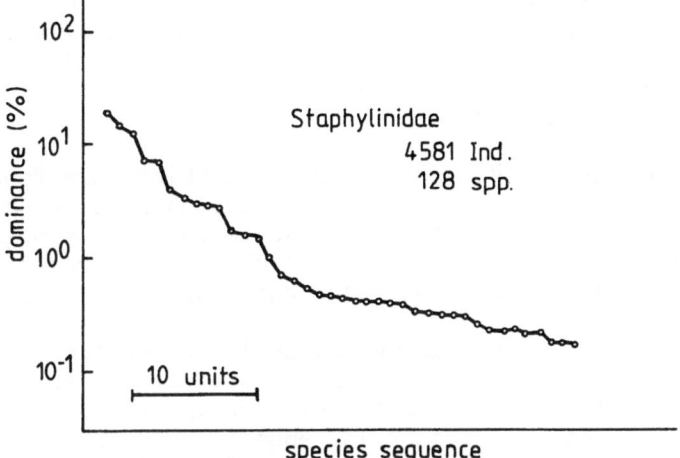

Fig. 1. Patterns of species abundance of staphylinids in an inland dune system of ca. 100x200 m surrounded by bogs and marshland

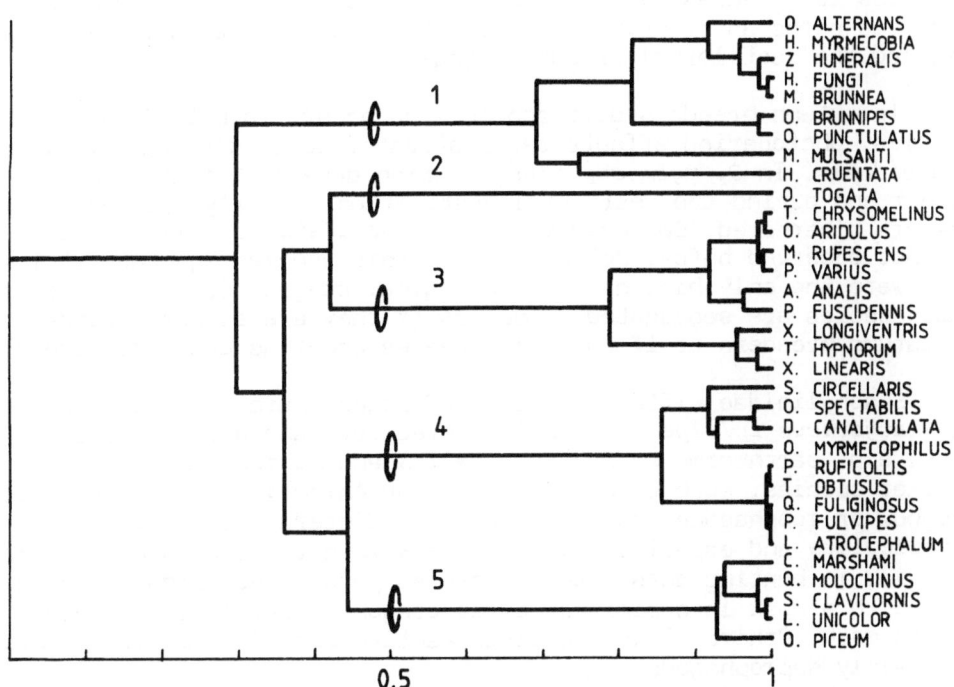

Fig. 2. Dendrogram (niche concerning the horizontal distribution pattern of 33 most common staphylinids within the inland dune system.

which is not unique but occurs in parallel samples of this zoogeo-
graphic region. Thus, faunas in particular sites do not seem to be
random assemblages with respect to structural habitat type. In this
paper, it is assumed that species will differ in habitat preference
if the overlap is less than 0.5.

It can be proved that not all species which were found within
this dune area are indigenous. Seven of the 33 species (*Atheta fungi*,
Philonthus spp. and *Tachyporus* spp.) leave the nearby moist agri-
cultural system during pre-diapause flights and search for overwintering sites in the dry and sandy dune soils.

Investigations on developmental cycles under different regimes
in the laboratory suggested that annual activity in staphylinids
is an important segregation mechanism too. Species do not only
oviposit during spring but also lay eggs in late summer and autumn
up untill October (Fig. 3). The annual life cycles of the species
as well as induction and termination of adult diapause mostly is
influenced by external factors such as temperature and photoperiod.
But also endogenous factors can be important (Topp, 1979b). Endogenous factors determine differences in *Tachyporus hypnorum* and
T. obtusus and result in a 4 weeks' displaced post-diapause flight
in these two species. When *T. obtusus* starts to oviposit within
the same sites as *T. hypnorum*, the latter species has already
reached the last larval or pupal stage.

In autumn-breeders differences also occur. In both *Othius*
species, photoperiod affects termination of adult diapause in most
individuals. In *O. myrmecophilus*, ovarian development starts in
August and during the years of investigation, usually the 2nd or 3rd
instars hibernated. Conversely, in *O. punctulatus* ovarian development
was not observed before October and in this species egg and 1st instar were the only hibernating preimaginal stages. It is assumed
that species are segregated by season if they are either spring-
or autumn-breeders or if such differences occur as described above.

Staphylinidae differ in food preference. Large species, such
as *Platydracus fulvipes*, are mainly predatory and attacked large
prey, e.g. earthworms *Lumbricus rubellus* and *Dendrobaena tenuis*.
Smaller species, such as *Atheta fungi* and *Amischa analis*, are saprophagous or zoophagous and attacked small diptera larvae. *Drusilla
canaliculata* and especially *Zyras humeralis* have developed specific
behavior in hunting ants. *Stenus* species pursue and grasp escaping
prey with a mask in a similar way as dragonfly larvae with their
labial mask. Others, like the winter-active *Lathrimaeum atrocephalum*,
and mainly saprophagous.

In the following, it is assumed that species differ in food
preference, when they are either predominantly saprophagous or zoophagous. Further, predators are separated, if morphological in-

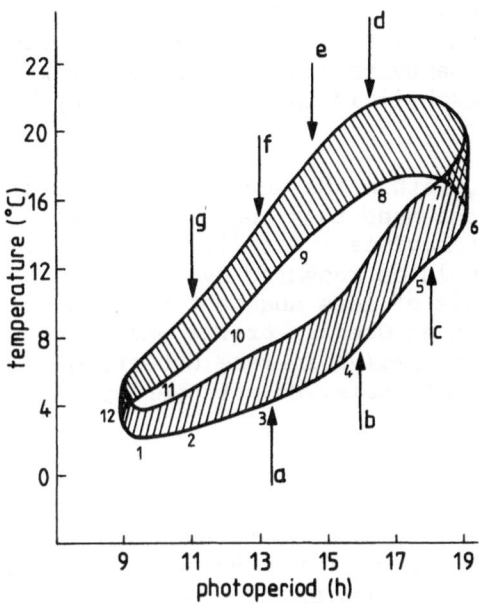

Fig. 3. Seasonal time pattern for the start of oviposition for most
abundant staphylinids of an inland dune system.
a) *Xantholinus linearis*, b) *Atheta fungi, Philonthus fusci-
pennis, Philonthus varius, Zyras humeralis, Drusilla canaliculata, Tachyporus hypnorum, Tachyporus chrysomelinus*,
c) *Tachyporus obtusus*, d) *Othius myrmecophilus*, e) *Olophrum
piceum, Lathrimaeum unicolor*, f) *Lathrimaeum atrocephalum*,
g) *Othius punctulatus*. The lines indicate monthly (1-12)
mean and maximum temperatures as well as the annual changes
of photoperiod including twilight time for the investigated
dune system.

dicators, such as size, are available. Segregation was evaluated
when size differences of at least a ratio of 1.33, were measured.
This procedure is appropriate for estimating significant differences.
In staphylinids size differences in addition gives the opportunity
for vertical segregation within the same sites. Small species are
not only active on the soil surface but enter the soil and use
small and narrow shelters for defence, food or a mate. In the dune,
several specimens of the tiny *Amischa analis* were found in the C-
horizon up to 25 cm below the surface.

The niche dimensions, mentioned so far, are not the only ones
which may keep species separated from each other. Diel time
partitioning may occur. Some colored species like *Tachyporus hypnorum*
are diurnal and dark ones like *Xantholinus linearis* are for the
most part nocturnal and usually hidden in crevices during the day.

Different foraging strategies, for example: direct and visual
hunting, random searching or hunting from an observation post or
specialization in behavior like hunting ants, may reduce direct
contact between staphylinid species and/or may be successful in
partitioning of prey.

Additionally, within the dunes horizontal distribution pattern
was still more pronounced than suggested above. For example, *Sipalia
circellaris* and *Pachygluta ruficollis* showed a more distinct
separation than could be measured by the above mentioned methods
and also individuals of less abundant species occurring only within
a single site, such as *Quedius brevis* which are raptorial in nests
of *Formica rufa* and *Quedius invreae* occurring in nests of the
Sphecidae *Mellinus aevensis*, do not overlap.

DISCUSSION

Niche dimensions as described above are ranked in a matrix comparing all possible pairs of the species of interest. For the indigenous species, only in one species pair (i.e. *Xantholinus linearis-
X. longiventris*) was no segregation mechanism detectable. In a second
species pair, whose members annually invade into the dune (i.e.
Tachyporus hypnorum-T. chrysomelinus), also no obvious separation
was found. Many species pairs differed in one dimension (n=53), but
most (n=174) differed in at least two niche dimensions (Fig. 4).
This observed pattern of the most common species significantly
deviates from a random distribution (P < 0.001) obtained by chi-
square test. Although there are several shortcomings as a first
step, this test may be used (Bowers and Brown, 1982).

When common species and rare species were analyzed together,

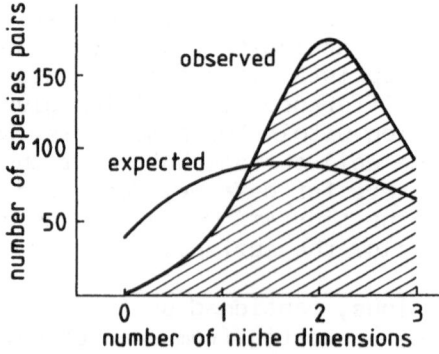

Fig. 4. Observed and expected segregation of staphylinids within an
        inland dune system concerning the niche dimensions: habitat,
        time and food (For abstraction smooth curves are sketched).

no such overdispersion of niches and deviation from random distribution (P > 0.05) occurred. Observation within the *Carex arenaria-
Rhacomitrium canescens* zone showed that the non-random pattern will
disappear if habitat differences of species are not taken into
account, although they exist in nature as seem for *Quedius brevis*
and *Qu. invreae* and/or if accidental invaders come in, and they do
not find any appropriate requisites in the area of immigration. In

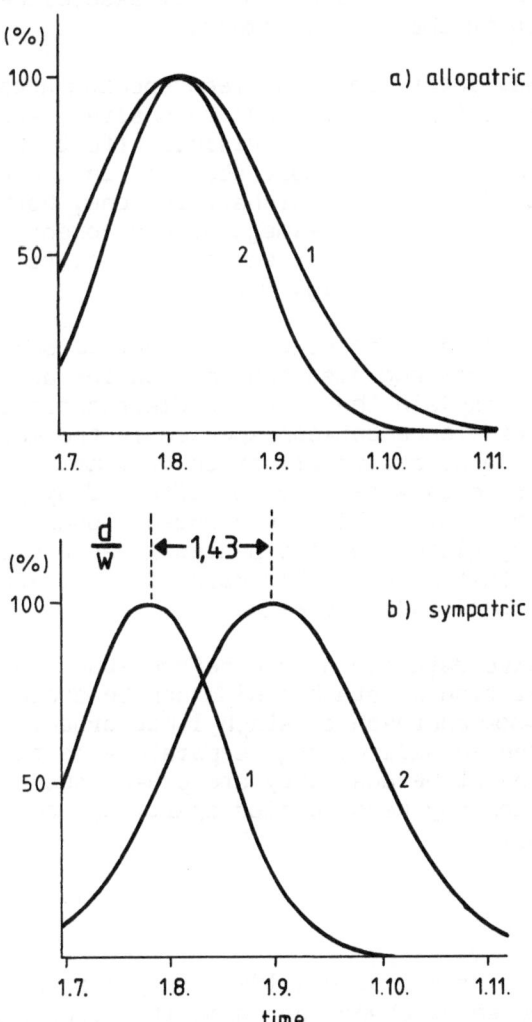

Fig. 5. Seasonal activity of 1. *Zyras humeralis* and 2. *Drusilla
canaliculata* regarded as Gaussian curves when occurring
allopatrically and sympatrically. d/w indicates the May-
index (see taxt).

this study, this holds true for *Phloeocharis subtilissima* and
Oxytelus rugosus within the driest sites.

Segregation of species as demonstrated, is not sufficient to
imply interspecific competition. Indeed, observations in the field,
as well as rearings in the laboratory, indicate the excellent adaptations of a species to its specific niche and from this it seems to
be unlikely that competition including both exploitation and interference, may cause co-evolutionary habitat shift or character displacement. Species of the above mentioned assemblages are quite
well pre-disposed to their environment.

Nevertheless, interference between species such as fighting
among larvae rather than adults, can be observed and in one species
pair interference seems to have a visible effect. *Zyras humeralis*
and *Drusilla canaliculata* are most abundant in two different habitat
assemblages (Fig. 2). But in a transition zone, both are the most
abundant staphylinids with pre-imaginal develoment during summer.
Occuring sympatrically in this transition zone, time partitioning
of both competitors occurs (Fig. 5).

Niche shift of the smaller *D. canaliculata* apparently is not
due to abiotic factors and also characteristics of habitats are
not sufficient to explain the observed distributions. Also, if it
is true that interference between species is the most important
factor, this has no selective effect on *D. canaliculata*. In both
species, oviposition in spring is not affected by pre-diapause
activity during autumn. The only consequence under constant environmental conditions, disregarding repeated invasions of the smaller
species, will be that *Z. humeralis* maintains its area of distribution whereas *D. canaliculata* does not.

More extensive data are needed before sound conclusion about
community organization in staphylinids can be drawn. Until now,
the above mentioned approach on staphylinid organization seems to
indicate that species will diverge separately as they evolve and
that species co-exist because they are pre-disposed to different
niches. Competition may keep species apart, an idea supported also
by Connell (1980).

REFERENCES

Bowers, M. A., and Brown, J. H., 1982, Body size and co-existence
    in desert rodents: chance or community structure, Ecology,
    63:391-400.
Connell, J. H., 1980, Diversity and the coevolution of competitors,
    or the ghost of competition past, Oikos, 35:131-138.
Connor, E. F., and Simberloff, D., 1979, The assembly of species
    communities: chance or competition? Ecology, 60:1132-1140.

Gauch, H. G., and Chase, G. B., 1974, Fitting the Gaussian curve to
    ecological data, Ecology, 55:1377-1381.
Gilpin, M. E., and Diamond, J. M., 1982, Factors contributing to
    non-randomness in species co-occurrences on islands, Oecologia
    (Berl), 52:75-84.
MacArthur, R. H., and Levins, R., 1967, The limiting similarity,
    convergence and divergence of co-existing species, Amer. Natur.,
    101:377-385.
May, R. M., 1976, Models for two interacting populations, in:"Theo-
    retical Ecology," 2nd edn., R.M. May, ed., Saunders, Philadel-
    phia.
Pianka, E. R., 1974, Niche overlap and diffuse competition, Proc.
    Nat. Acad. Sci. U.S.A., 71:2141-2145.
Simberloff, D., and Boeckler, W., 1981, Santa Rosalia reconsidered:
    size ratios and competition, Evolution, 35:1206-1228.
Sneath, P. H. A., and Sokal, R. R., 1973, "Numerical taxonomy,"
    Freeman, San Francisco.
Topp, W., 1979a, Verteilungsmuster epigäischer Arthropoden in einer
    Binnendünenlandschaft, Schr. Naturw. Ver. Schlesw.-Holst., 49:
    61-79.
Topp, W., 1979b, Vergleichende Dormanzuntersuchungen an Staphyliniden
    (Coleoptera), Zool. Jahrb. Syst., 106:1-49.

LOW TEMPERATURE INDUCED DIAPAUSE STILL EXTANT IN A TROPICAL

PARASITOID SPECIES

Y. Carton

Laboratoire de Génétique Evolutive
C.N.R.S.
Gif-sur-Yvette,91190, France

The discovery of a new species of *Leptopilina* named *Leptopilina boulardi* (Barbotin et al., 1979; Nordlander, 1980),a larval parasitoid of *Drosophila melanogaster*, in Guadeloupe Island (West Indies) led us to look for the possible existence of diapause in this tropical species.

In tropical countries, especially in the savannas and rainforests, physical environmental conditions are much more stable and, at first sight, resources should be available all the year round. On present evidence the existence of diapause in such countries seems quite uncommon (Denlinger, 1974).

We have demonstrated (Claret and Carton, 1980) the occurence of facultative diapause at the prepupal stage of this parasitic wasp. It is induced by a relatively low temperature (17.5°C) while photoperiod is absolutely ineffective (Fig. 1).

It is a genuine diapause, not simply a quiescence, because it always occurs at the same developmental stage (third instar larvae) after a period of growth from egg to prepupa at temperatures which induce diapause. Quiescence is a suppression of developmental processes that is imposed by adverse physical conditions; its duration is the same as that of the adversity. Duration of growth at 20°C is 26-27 days for individuals which do not undergo diapause whereas those which undergo diapause are still at the third larval instar after 60 days. In addition, 87% of the diapausing larvae moult and 13% are still at the same stage 40 days later, when placed at 25°C to ensure termination of diapause; in a quiescence all the larvae should metamorphose at 25°C within a much shorter time. (Carton and Claret, 1982).

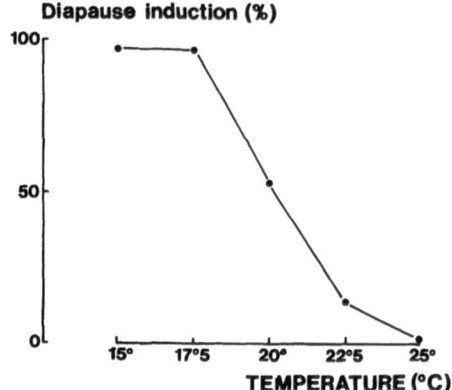

Fig. 1. Incidence of larval diapause in *L. boulardi* reared throughout larval life at different temperatures (with a daily 12 h photophase).

The population used for this study was collected in a lowland place, at 65 m elevation (Domaine Duclos, Basse Terre, Guadeloupe).

The existence of such a diapause is more surprising since the thermal conditions causing diapause to occur are not met in this area where the animals were collected. Consequently it could appear that this population never enters diapause throughout the year. Non-adaptive diapause seems inconceivable; this complex physiological mechanism could hardly have lasted in the absence of a selective mechanism. This mechanism is much more likely to have a definite ecological meaning in some areas of Guadeloupe at least.

During 1980 and 1982's years, the author was able to trap this parasitic wasp at several different places on Guadeloupe Island, located at different elevations (Fig. 2):
- Maison d'hôte (Petit Bourg)........................ 25 meters
- INRA refuge (Domaine Duclos)...................... 65 meters
- Rivière Vieux Fort................................. 222 meters
- Dos d'Ane... 452 meters
- Morne à Louis..................................... 743 meters

Meteorological records in these mountains show that temperatures are lower in high elevations (Fig. 3). From the diapause curve, we can estimate, for each temperature, the percentage of diapausing larvae. We can observe in a station with an elevation of 423 m a period of four months during which the mean temperature is around 21°C and would correspond to a diapause induction of 35%-45% approximately. At the highest station (743 m), mean temperatures of 19°-20°C are observed for five months during which diapause induction varies from 45% to 75%.

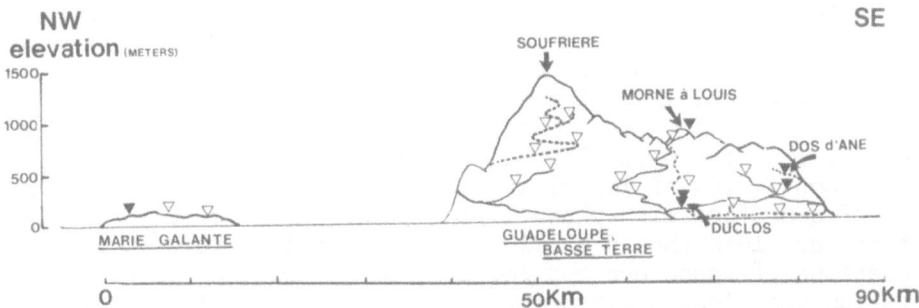

Fig. 2. Elevations (meters) and relative distance (kilometers) of
the 24 sampling sites investigated for *Drosophila* and parasitic wasp (*L. boulardi*) in Guadeloupe and Marie Galante
islands. Every site yielded *Drosophila*; dark triangles indicate those sites where both parasitic wasp and flies are
present and open triangles those where *L. boulardi* is lacking.

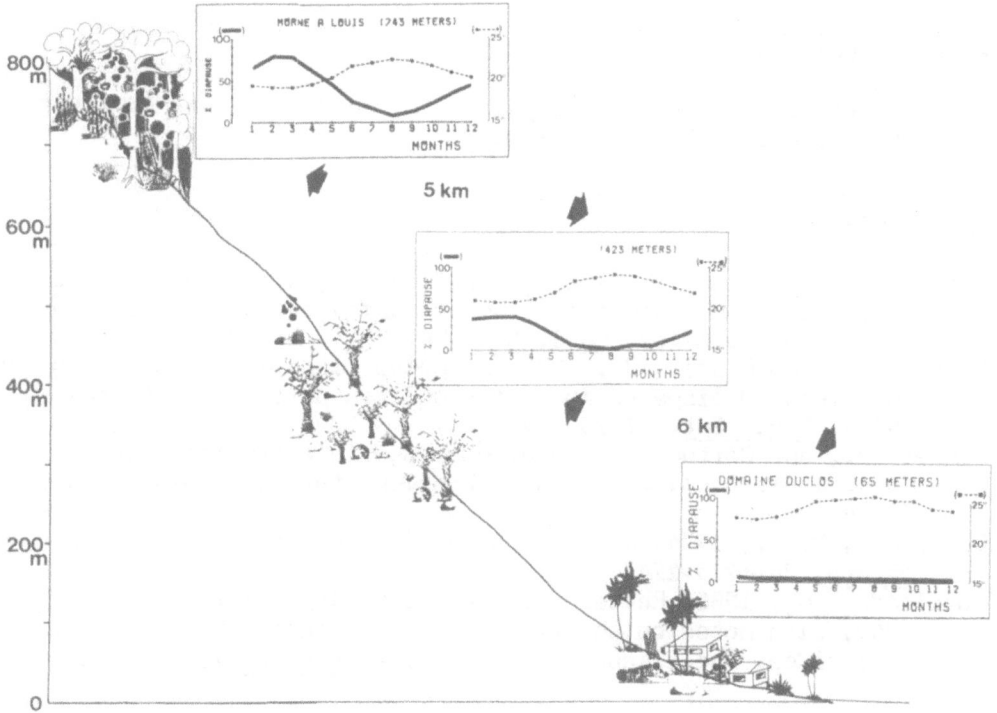

Fig. 3. Annual variation in diapause frequency (solid line) in three
elevation -characteristic habitats in Guadeloupe Island
(dotted line indicates annual temperature variation).

At these altitudes, thermal conditions are less favorable for host development; we may suppose that such a period of low temperatures should correspond to a quantitative decrease of *D. melanogaster* population. Thus, the capability of *L. boulardi* to enter diapause grants it an advantage and acts as a strategy against a temporary lack of hosts. Secondly, the preservation of diapause in populations living at low altitude would, however, mean the maintenance of gene flow with those living at higher altitudes. These stations are less than 5 or 11 km away from "INRA refuse" station. Populations of wasps are not isolated and a certain amount of migration must inevitably occur; as a matter of fact, we may suppose that the populations are contiguous and that the exchanges of individuals may well be quite high.

We have also discovered this parasite in Marie Galante Island (Fig. 2) which is a lowland place (highest elevation: 154 meters). It will be very interesting to perform experiments with this population to look for absence of diapause. If not, our hypothesis might be revised, unless we suppose a migration of parasites between the two islands.

ACKNOWLEDGEMENTS

Support of CNRS (ATP ecophysiologie, no 51-3571 and ATP CNRS-NSF, USA) is gratefully acknowledged.

REFERENCES

Barbotin, F., Carton, Y., and Kelner-Pillault, S., 1979, Morphologie et biologie de *Cothonaspis* (*Cothonaspis*) *boulardi* n. sp., parasite de Drosophiles, Bull. Soc. Ent. Fr., 84:19-24.

Carton, Y., and Claret, J., 1982, Adaptive significance of a temperature induced diapause in a cosmopolitan parasitoid of *Drosophila*, Ecol. Ent., 7:239-247.

Claret, J., and Carton, Y., 1980, Diapause in a tropical species *Cothonaspis boulardi* (Parasitic hymenoptera), Oecologia, 45: 32-34.

Denlinger, D. L., 1974, Diapause potential in tropical flesh fly, Nature, 252:223-224.

Nordlander, G., 1980, Revision of the genus *Leptopilina* Förster, 1969, with notes on the status of some genera (Hymenoptera, Gynipoidea, Eucoilidae), Entomologica Scandinavica, 11:428-453.

COLD TOLERANCE IN CANADIAN ARCTIC INSECTS

Richard A. Ring

Biology Department
University of Victoria
Victoria, B.C. V8W 2Y2, Canada

INTRODUCTION

Canada is ideally located for the comparative study of insect cold tolerance mechanisms due to its wide range in latitudes, altitudes and climatic zones. Although the high altitude insects have not been enumerated the arctic insect fauna is rich and varied, the current estimate being 1640 identified species, including Collembola and Protura (Danks, 1981). This includes representatives from about 16 different orders, thus the diversity of taxa makes the arctic region also suitable for phylogenetic studies. Table 1 is a compilation of the available data on identified species in various comparable regions of the world where low winter tempreratures and short growing seasons prevail.

Table 1. Insects in some representative areas of high latitude or altitude

Region	Number of Species
Arctic[a]	1640
Hazen Lake, Ellesmere Island(81°49′N)	c. 300
Ellef Ringnes Island (78°47′N)	37
Queen Elizabeth Islands (c. 76°90′N)	399
Bathurst Island (75°43′N)	71
Devon Island (75°40′N)	156
Barrow, Alaska (71°20′N)	232
Greenland (60°84′N)	c. 600
Hardangervidda, Norway (60°37′N)	c. 313
Antarctica	63

a: Canada and Alaska, from the treeline north

Conventional knowledge of insect cold tolerance at the present time recognizes two major overwintering strategies: freezing tolerance, the ability to survive actual ice formation within the body and freezing intolerance (or susceptibility),the ability to avoid freezing by supercooling to extreme, but varying, degrees. Insects within each category share many important biological and biochemical features, but evidence is accumulating to show that many anomalies occur. Such variations indicate that cold tolerance mechanisms have evolved independently on a number of occasions.

This paper presents the results of a comparative study of some insects from high latitude, high altitude and a more temperate area of Canada. It will be seen that all species conform to one or other of the two major overwintering strategies but that there are many specific variations within each general theme.

MATERIALS AND METHODS

Insects

The cold-hardy insects used in this study were collected in the tundra of the N.W.T. (69°35'N), the MacKenzie River Delta area (68°-69°N), the Rocky Mountains (51°N) at elevations over 2000 m, or in the Victoria, B.C. area (48°N). Table 2 presents a list of these insects, their distribution and habitat.

Low Temperature Acclimation

All Specimens were returned to the laboratory and placed in acclimation chambers at an initial temperature of 8°C. They were held at this temperature without additional food for about 4 weeks ("preacclimated" insects) before being submitted to the following acclimation regimen:- approximately 16 weeks during which the temperatures were gradually reduced to -15° or -17°C in decrements of 5°-6°C. Rapid de-acclimation was accomplished by warming the insects at 22°C for 72 hours. In order to minimize desiccation of specimens they were maintained in perforated canisters (a) with bark under which they were originally found, (b) with some leaves and twigs from their host plants, or (c) within their galls. Approximately every third week a sample of enough individuals to provide 100-250 mg fresh weight was removed from the acclimation chamber, weighed, tasted for supercooling capacity and freezing tolerance, and finally subjected to biochemical analysis.

Determination of Supercooling Points and Freezing Tolerance

Supercooling points were measured with a 30 gauge copper-constantan thermocouple attached to a recording potentiometer at a cooling rate of 1°-2° per minute (Ring and Tesar, 1980).

Determination of freezing tolerance was made according to
Asahina (1969), Sømme (1974),and Zachariassen and Hammel (1976),
the major criterion of tolerance being the ability to survive and
perform coordinated movements after freezing at the supercooling
point or at a lower temperature.

Biochemical Analyses

The determination of accumulated polyols along with any concurrent changes in glycogen and blood sugar levels was achieved
essentially by the methods of Van Handel (1965),and Kartnig and
Wegschaider (1971) as modified by Ring and Tesar (1980). Separation
of polyols and sugars was achieved by thin layer chromatography
(TLC), and their quantification by a colorimetric technique using
a Unicam SP 8000 scanning spectrophotometer. With these methods
trehalose, sucrose, glucose, fructose and glycerol with their
respective Rf values of 0.43, 0.50, 0.56, 0.60 and 0.71 were
identified. There were also traces of two unidentified compounds
with Rf values of 0.73 and 0.76. Glycogen was measured by the
anthrone reagent assay, while total amino acid content and lipid
analysis, where carried out, were based on the colorimetric
techniques outlined by Ring and Tesar (1980). Water content was
determined by substracting the dry weight, obtained after 24 hours
in the freeze drier, from the fresh weight of the sample.

RESULTS AND DISCUSSION

Nine species of insects from three different geographical
regions of Canada were examined for freezing tolerance, supercooling capacity, water content and changes in biochemical characteristics during acclimation to subzero temperatures for varying
periods of time. Six species proved to be freezing tolerant (*Pytho
americanus* and *P. deplanatus* (Coleoptera, Pythidae (Salpingidae));
Xylophagus sp. (Diptera, Xylophagidae); two species of *Pontania*
(Hymenoptera, Tenthredinidae); and *Gynaephora* sp. (Lepidoptera,
Lymantriidae) , while the remaining three species were freezing
susceptible *Coccinella quinquenotata* (Coleoptera, Coccinellidae);
Rhabdophaga sp. (Diptera, Cecidomyiidae); and *Diastrophus kincaidii*
(Hymenoptera, Cynipidae) . The geographic distribution of the
species gave no indication of whether it was freezing tolerant or
freezing susceptible since species from all three regions (arctic,
alpine and southwest B.C.) were found in each category.

The majority of insects adhered to the generally accepted profiles of overwintering response. In the freezing tolerant species
supercooling points were relatively high (mean = -8^{0}C), moderate
amounts of polyols were accumulated (mean= 9% FW), and water levels
declined in a uniform manner during low temperature acclimation.
Glycogen levels in all species also declined significantly during
this period, indicating that this reserve is the most likely source

Table 2. Insects used in the study

Species	Stage	Locality	Habitat
FREEZING TOLERANT			
Coleoptera			
1 _Pytho americanus_	(a) Adults (b) Larvae	MacKenzie River Delta, N.W.T. (68°43'N)	Under bark of fallen trees
Pytho americanus	Larvae	Boom Lake, Rocky Mts. (51°16'N); elevation over 2000 m	Under bark of fallen trees
3 _Pytho deplanatus_	Larvae	Boom Lake, Rocky Mts. (51°16'N); elevation over 2000 m	Under bark of fallen trees
Diptera			
4 _Xylophagus_ sp.	Larvae	MacKenzie River Delta, N.W.T. (68°43'N)	Under bark of fallen trees
Hymenoptera			
5 _Pontania_ sp. 1	Larvae	Tundra, near Tuktoyaktuk, N.W.T. (69°35'N)	Leaf galls on _Salix glauca_
Pontania sp. 2	Larvae	Victoria, B.C. (48°25'N)	Leaf galls on _Salix lasiandra_

Lepidoptera

7	*Gynaephora* sp.	Larvae	Tundra, near Tuktoyaktuk, N.W.T. (69°35'N)	Feeding on willows

FREEZING SUSCEPTIBLE

Coleoptera

8	*Coccinella quinquenotata*	Adults	Inuvik, N.W.T. (68°32'N)	Aphid-infested willows

Diptera

9	*Rhabdophaga* sp.	Larvae	MacKenzie River Delta N.W.T. (68°40'N)	Stem galls on *Salix lanata*

Hymenoptera

10	*Diastrophus kincaidii*	Larvae	Victoria, B.C. (48°25'N)	Thimble Berry, *Rubus parviflorus*

for polyol and sugar synthesis. Statistical analyses of the data obtained from *P. americanus* larvae (Ring and Tesar, 1980) has already shown the very high correlation coefficients between increasing glycerol and decreasing glycogen concentrations. Another possible source of metabolically inexpensive glycerol, however, could come from glyceride reserves (Baust and Miller, 1972; Morrissey and Baust, 1976), and a recent study by Yaginuma and Yamashita (1980) has shown that in diapausing eggs of *Bombyx mori* a small proportion (33%) of free glycerol is derived from lipid glycerol while the remainder comes from glycogen. The results from *P. americanus* (arctic population), the only species investigated in this respect, tend to corroborate this finding (Table 3).

Total lipid content declined during low temperature acclimation by factors of 2.4 and 1.7 in adults and larvae respectively. This was due mainly to decreases in fatty acids, diglycerides and triglycerides.

Many specific variations in freezing tolerance were apparent among the six test species. *P. americanus* (arctic population) was freezing tolerant in both larval and adult stages, a condition which would be congruous with its northern distribution and allow it to spread its life cycle over a number of growing seasons (Downes, 1962, 1964, 1965; MacLean, 1975, Danks, 1978; Ring and Tesar, 1980). Both stages had almost identical biochemical characteristics and the same degree of freezing tolerance. Although glycerol was the most common and abundant polyol synthesized in the freezing tolerant insects, it was replaced by sorbitol in *Pontania* spp. (Table 7). Only one polyol was synthesized in any one species (although recent studies using HPLC analyses have demonstrated the presence of small amounts of erythritol in both

Table 3. Total lipid content during acclimation of *P. americanus*

Acclimation (weeks)	Total lipids* (% FW)	
	ADULTS	LARVAE
2	9.08	4.48
4	8.92	4.19
6	-	-
8	-	-
10	6.94	2.51
12	6.11	3.20
14	5.65	2.76
16	4.28	2.41
18	3.80	3.17

*fatty acids, diglycerides, triglycerides and phospholipids

P. americanus and *P. deplanatus* (Ring, unpublished data)).In marked
contrast to the above species, *Xylophagus* larvae did not accumulate
any polyols or blood sugars, but was tolerant to freezing down to
at least -43°C (Ring, 1981) (Table 8). The supercooling points
throughout low temperature acclimation remained relatively high
(-4°C to -6°C). *Xylophagus* shares these characteristics with only
a few other freezing tolerant insects (Sømme, 1964; Zachariassen
et al., 1979). Perhaps the most interesting variation within this
group was found in *P. deplanatus* (Table 4). In addition to having
moderate levels of glycerol and the ability to tolerate freezing
down to at least -55°C, this species had extremely low supercooling
points (-54°C), indeed the lowest ever recorded for this category.
The most feasible explanation for the co-existence of freezing
tolerance and low supercooling points is that *P. deplanatus* is
physiologically and biochemically incapable of synthesizing haemo-
lymph nucleating agents and that nucleators in the gut and other
fluid compartments of the body are absent in overwintering larvae.
The adaptive significance of this strategy is that larvae would
normally overwinter in a supercooled state, but if freezing did
occur (e.g. by inoculation from contact water in the hibernaculum)
then the antifreeze function of glycerol and blood sugars would
assume a cryoprotective role (Ring, 1982). *P. deplanatus* is not
unique in this respect, and a literature search shows a number of
other insects which are freezing tolerant but with concurrent low
supercooling points (Table 5). The water content of *P. deplanatus*
was reduced by more than one half during the 14 week exposure to
subzero temperature (Table 6). There is evidence from some other
freezing tolerant insects, however, that much of this water is not

Table 4. Variations between two congeneric species of freezing
        tolerant beetle larvae

		Pytho americanus		*Pytho deplanatus*
Collection site		Arctic	Alpine	Alpine
Supercooling point* (°C)		-7.1	-6.0	-54.0
Glycerol content* (% FW)		8.20	6.70	13.20
Glycogen content (% FW)	A.	4.20	5.60	4.10
	B.	0.17	0.14	0.20
Total sugar content (% FW)	A.	2.61	3.32	1.50
	B.	0.46	1.15	5.50

A. pre-acclimation level
B. level at peak of acclimation

*all measurements in this and subsequent Tables were those taken
 at the peak of low temperature acclimation.

Table 5. Freezing tolerant insects with low supercooling points

Insect	Stage	Supercooling Point (°C)*	Cryoprotectant (%)	Reference
Coleoptera				
Pytho deplanatus	Larva	-54	Glycerol (13.2%)	Ring, Tesar (1981)
Mordellistena unicolor	Larva	-26	?	Baust et al. (1979)
Lepidoptera				
Plygonia sp.	Adult	-25	Glycerol (9%)	Miller (1978)
Martyrhilda ciniflonella	Adult	-23	Glycerol (8%)	Miller (1978)
Diptera				
Mycetophilia sp.	Adult	-33	Glycerol (14%)	Miller (1978)
Exechia sp.	Adult	-33	Glycerol (14%)	Miller (1978)
Hymenoptera				
Bracon cephi	Larva	-47	Glycerol (29%)	Salt (1959)
Eurytoma gigantea	Larva	-49	Glycerol (23.4%)	Sømme (1964)
"	Larva	-27	?	Baust et al. (1979)
Eurytoma obtusiventris	Pupa	-29	?	Baust et al. (1979)
Rogas sp.	Adult	-30	Glycerol (?)	Miller (1978)

* lowest supercooling point recorded in winter or during acclimation in the laboratory

Table 6. Similarities between freezing tolerant beetle larvae and
freezing intolerant stem-gall dipteran larvae

		Pytho deplanatus (freezing tolerant)	*Rhabdophaga* sp. (freezing intolerant)
Collection site		Alpine	Arctic
Supercooling point (^{o}C)		-54.0	-61.6
Glycerol content (% FW)		13.20	20.10
Glycogen content (% FW)	A.	4.10	13.60
	B.	0.20	0.70
Total sugar content (% FW)	A.	1.50	5.30
	B.	5.50	2.20
Relative H_2O content*		0.57	0.45

*the highest level of hydration in the de-acclimated state has been
assigned the value of 1.00

lost to the insect but held in a metabolically "bound" state (Ring
and Tesar, 1980, 1981). Storey et al. (1981) suggest that layers
of "bound" water around intracellular macromolecular structures
serve to limit intracellular freezing and protect molecules such
as proteins from denaturation. Furthermore, as the level of bulk
water which acts as a solvent would be expected to increase,
resulting in further depression of supercooling points (assuming
absence of nucleating agents)(Baust et al., 1979). In the freezing
susceptible species supercooling points were low to very low.
Rhabdophaga larvae, which overwinter within stem galls and are
exposed to ambient air temperatures during the long arctic winter,
exhibited extreme depression of supercooling points down to -61.6^{o}C
in some individuals (Ring and Tesar, 1981) (Table 6). This is the
lowest super-cooling point ever recorded for an insect, although
Miller and Werner (1980),and Miller (1982) have recently found
similar extensions in the supercooling range of willow gall insects
from Alaska. Glycerol levels of over 20% and glycogen content of
over 13% were also the highest recorded among the rest species.
Extractable water fell by almost one half, as in *P. deplanatus*,
indicating, perhaps that enhanced supercooling capabilities can
only occur in living systems in which a large proportion of water
is in a metabolically bound state.

   Diastrophus kincaidii had similar though less pronounced de-
pressions in supercooling points and increases in concentrations,
reflecting its more southerly distribution and a more temperate
winter climate in southern coastal B.C. (Table 7). One feature
that this species and *Rhabdophaga* had in common was the presence
of relatively low supercooling points (-22°C) in the non-acclimated,

Table 7. Variations among some southern and arctic leaf-gall hymenopteran larvae

		Pontania sp. 1 (freezing tolerant)	*Pontania* sp. 2 (freezing tolerant)	*Diastrophus kincaidii* (freezing intolerant)
Collection site		Arctic	Victoria, B.C.	Victoria, B.C.
Supercooling point (°C)		-9.0	-7.5	-33.0
Glycerol content (% FW)		sorbitol	sorbitol	3.80
Glycogen content (% FW)	A.	2.67	1.60	2.70
	B.	0.45	0.10	0.30
Total sugar content (% FW)	A.	2.04	0.90	2.20
	B.	5.60*	4.00*	1.80

* includes small amounts of sucrose

Table 8. Some anomalous insects

		Xylophagus sp. (larvae) Diptera (freezing tolerant)	*Coccinella quinquenotata* (adults) Coleoptera (freezing intolerant)	*Coleomegilla maculata* (adults) Coleoptera (freezing intolerant)
Collection site		Arctic	Arctic	W. New York State
Supercooling point (°C)		-6.0	-24.0	-18.4
Glycerol content (% FW)		-	-	-
Glycogen content (% FW)	A.	1.10	1.30	Not detrmined
	B.	0.07	0.56	
Total sugar content (% FW)	A.	1.22	0.44	Traces of trehalose
	B.	1.28	1.54*	0.80
Relative H$_2$0 content		0.90	0.87	

*includes relatively large amounts of sucrose

summer stages. This implies that some freezing susceptible insects are cold hardy even before exposure to low temperatures, and that some degree of cold tolerance can occur in the absence of glycerol. This statement is certainly true in the case of *Coccinella quinquenotata* which overwinters in the adult stage (Table 8). No glycerol or other polyol was synthesized during 17 weeks of low temperature acclimation, while supercooling points were lowered from $-8^{\circ}C$ in the non-acclimated insects to $-24^{\circ}C$. Glycogen, which was low in concentration to begin with (1.3% FW), declined by a factor of only 2.3, the least of all species in the study. Total blood sugar content, however, increased by a factor of 3.5 from 0.44 to 1.54% FW, in marked contrast to the other freezing susceptible species which showed significant decreases in sugars during acclimation.

Coccinella quinquenotata has almost identical overwintering properties to those of another freezing susceptible coccinellid, *Coleomegilla maculata* (Baust and Morrissey, 1975). These beetles differ so widely from the accepted profile of overwintering response in this category that further studies will have to be conducted to elucidate the mechanisms of this unique strategy. Perhaps amino acids play a role in cold hardiness in these species, since it has been shown that increases in amino acids contribute to the cold tolerance of *Pytho americanus* (Ring and Tesar, 1980), *Ips typographus* (Hansen et al., 1980), and *Mytilus edulis* (Williams, 1970).

The major conclusion that can be drawn from this study is that different species have different adaptations to low temperature survival even if they fall within one or other of the two major overwintering strategies. It seems that the physiological and biochemical basis of cold hardiness in insects is based on a number of interacting factors, although the correct combination of such factors and complexity of the process are still obscure (Baust, 1981). The marked differences that exist in the overwintering responses of some species that are closely related phylogenetically would indicate that cold tolerance mechanisms have evolved independently on a number of different occasions. The contentions of some authors that (i) only species which hibernate predominantly above the snow level have the ability to accumulate polyols in winter and (ii) freezing tolerance develops particularly in species hibernating in extremely cold environments, are not supported by these results. Further research into the role of "bound" water, the presence or absence of nucleating agents and thermal hysteresis proteins, and the comparative biochemistry of carbohydrate and lipid metabolism should help elucidate the basic mechanisms of cold tolerance in insects.

ACKNOWLEDGEMENTS

This research was supported by an operating grant from NSERC

of Canada and a Faculty Research Grant from the University of
Victoria. I would like to thank Mr. D. Tesar for technical assistance;
Dr. D. Sherstone and Mr. John Ostrick of the Department of Indian
and Northern Affairs, Inuvik, and Dr. G. D. Hobson and Mr. E. Chap-
man of the Department of Energy, Mines and Resources Canada, Tukto-
yaktuk, for the facilities placed at my disposal in the N.W.T.

REFERENCES

Asahina, E., 1969, Frost resistance in insects, in: "Advances in
    insect physiology," J.W.L. Beament, J.E. Treherne, and V.B.
    Wigglesworth, eds., Vol. 6, Academic Press, New York.
Baust, J. G., 1981, Biochemical correlates to cold hardening in
    insects, Cryobiol. 18:186-198.
Baust, J. G., and Miller, L. K., 1972, Influence of low temperature
    acclimation on cold hardiness in Pterostichus brevicornis,
    J. Insect Physiol., 18:1935-1947.
Baust, J. G., and Morrissey, R. E., 1975, Supercooling phenomenon
    and water content independence in the overwintering beetle,
    Coleomegilla maculata, J. Insect Physiol., 21:1751-1754.
Baust, J. G., Grandee, R., Condon, G., and Morrissey, R., 1979, The
    diversity of overwintering strategies utilized by separate
    populations of gall insects, Physiol. Zool., 52:572-580.
Block, W., and Zettel, J., 1980, Cold hardiness of some Alpine
    Collembola, Ecol. Ent., 5:1-9.
Danks, H. V., 1978, Modes of seasonal adaptation in the insects, I,
    Winter survival, Can. Ent., 110:1167-1205.
Danks, H. V., 1981, Arctic arthropods: a review of systematics and
    ecology with particular reference to the North American fauna,
    Entomological Society of Canada, Ottawa.
Downes, J. A., 1962, What is an arctic insect?, Can. Ent., 94:143-162.
Downes, J. A., 1964, Arctic insects and their environment, Can. Ent.,
    96:279-307.
Downes, J. A., 1965, Adaptations of insects in the arctic, Annu. Rev.
    Ent., 10:257-274.
Hansen, T. E., Viyk, M. O., and Luyk, A. K., 1980, Biochemical
    changes and cold-hardiness in overwintering bark-beetles Ips
    typographus, Ent. Rev. 59:9-12
Kartnig, T., and Wegscheider, O., 1971, Eine Möglichkeit zur Identi-
    fizierung von Zuckern auskleinsten Meugen von Glycosiden oder
    aus Zucker gemischen, J. Chromatogr., 61:375-377.
MacLean, S. F., Jr., 1975, Ecological adaptations of tundra in-
    vertebrates, in: "Physiological adaptation to the environment,"
    F. J. Vernberg, ed., Intext Educational Publishers, New York.
Miller, L. K., 1978, Physiological studies of arctic animals, Comp.
    Biochem. Physiol., 59A:327-334.
Miller, L. K., 1982, Variations in freezing tolerance in insects
    overwintering in interior Alaska, Cryo-Letters (in press).
Miller, L. K., and Werner, R., 1980, Suporcooling to -60°C: an

extreme example of freezing avoidance in northern willow gall
insects, Cryobiol., 17:621-622.

Morrissey, R., and Baust, J. G., 1976, The ontogeny of cold tolerance
in the gall fly, *Eurosta solidagensis*, J. Insect Physiol., 22:
431-437.

Ring, R. A., 1981, The physiology and biochemistry of cold tolerance
in arctic insects, J. therm. Biol., 6:219-229.

Ring, R. A., 1982, Freezing tolerant insects with low supercooling
points, Comp. Biochem. Physiol. A, (in press).

Ring, R. A., and Tesar, D., 1980, Cold hardiness of the Arctic beetle,
Pytho americanus Kirby (Coleoptera, Phythidae (Salpingidae)),
J. Insect Physiol., 26:763-774.

Ring, R. A., and Tesar, D., 1981, Adaptations to cold in Canadian
arctic insects, Cryobiol., 18:199-211.

Salt, R. W., 1959, Role of glycerol in the cold-hardening of *Bracon
cephi* (Gahan), Can. J. Zool., 37:59-69.

Sømme, L., 1964, Effects of glycerol on cold-hardiness in insects,
Can. J. Zool., 42:87-101.

Sømme, L., 1974, The overwintering of *Pelophila borealis* Payk III,
Freezing tolerance, Norw. J. Ent., 21:131-134.

Storey, K. B., Baust, J. G., and Buescher, P., 1981, Determination
of water "bound" by soluble subsellular components during low-
temperature acclimation in the gall fly larva, *Eurosta solidagensis*, Cryobiol., 18:315-321.

Van Handel, E., 1965, Microseparation of glycogen, sugars and lipids,
Analyt. Biochem., 11:266-271.

Williams, R. J., 1970, Freezing tolerance in *Mytilus edulis*, Comp.
Biochem. Physiol., 35:145-161.

Yaginuma, T., and Yamashita, O., 1980, The origin of free glycerol
accumulated in diapause eggs of *Bombyx mori*, Physiol. Ent., 5:
93-97.

Zachariassen, K. E., and Hammel, H. T., 1976, Freeze-tolerance in
adult tenebrionid beetles, Norw. J. Zool., 24:349-352.

Zachariassen, K. E., Hammel, H. T., and Schmidex, W., 1979, Studies
on freezing injuries in *Eleodes blanchardi* beetles, Comp.
Biochem. Physiol., 63A:199-202.

SEASONAL ACTIVITY OF SOIL FAUNA IN A PHRYGANIC (EAST MEDITERRANEAN) ECOSYSTEM

S. Sgardelis and N. S. Margaris

Laboratory of Ecology
Faculty of Physics and Mathematics
University of Thessaloniki
Thessaloniki, Greece

INTRODUCTION

Mediterranean type climate is characterized by an alternation of hot and dry summers to mild and wet winters, while the coincidence of summer drought with the high temperatures consists the most important factor for the living organisms in Mediterranean regions (Nahal, 1981).

Phrygana which appear at the dry end of the precipitation gradient of Mediterranean climate regions are characterized by seasonal dimorphism (Margaris and Vokou, 1982). This feature results to a seasonal pattern of leaf fall. In parallel seasonality which has been observed in the decomposers' activity (Fousseki and Margaris, 1981) is typical of most mediterran ecosystems (Shaefer, 1973).

In this work a preliminary description of the seasonal fluxes of soil fauna is provided.

SITE DESCRIPTION

The system under study is located in Attica, close to the village of Keratea, about 50 km SE of Athens. It is dominated by the seasonally dimorphic shrubs *Sarcopoterium spinosum*, *Genista acanthoclada*, and *Thymus capitatus*. The site is characterized by a great spatial heterogeneity with pathways separating shrubby patches, and a high number of stones on the ground.

Litter appears well restrained under the shrub canopy over-

31

laying a loose layer consisted mainly of animal excrements and
detritus. Typical humus layer does not exist.

Climatic data for the sampling period have been taken from the
meteorological station of Hellinikon airport located at a distance
of about 20 km.

METHODS

Samples of both Macro- and Mesofauna were taken in monthly
intervals. Macrofauna was hand-sorted, from five quadrats, 50x50 cm^2
and approximately 30 cm depth. Such a sampling may provide rough
estimations of density. During summer, due to the low densities
and high mobility of the dominant surface animals, only relative
abundances were estimated by searching in greated surfaces.

Soil mesofauna was extracted by the Berlese-Tülgren funnel
method from 5 samples per month, taken from the litter under the
shrubs by a core of 5 cm diameter.

Analysis of the data has been done by Reciprocal Averaging
(Hill, 1973) in order to detect the general trends of the evolution
of soil animals in time. R.A. is selected because of its duality in
ordinating variables to go with corresponding samples (Hill, 1973).

RESULTS AND DISCUSSION

Macrofauna

Our data are given in Table 1 where months and animal groups
are ranked according to R.A. first axis. Before analysis data were
transformed to percentages so that each month total accounts 100.
The dominant groups of the wet period were Julidae, Enchytraeidae
(although only large individuals were included), Insect larvae
and Lumbricidae, while Coleoptera, Thysanura and Araneae dominate
during summer. Most of the animals are found under the shrubs,
where litter accumulates, whereas certain groups aggregate under
stones (Lithobiomorpha, Araneae, Isopoda) and some may be found in
bare soil (Enchytraeidae, some larvae and Geophilomorpha).

As regards seasonality we can identify 3 distinct periods
(Fig. 1) at the 3 first axes of R.A. A wet period (I) from the
first rains of autumn to February, a moderate period (II) from
March to May and the period of drought (III).

The factor possibly related to such groupings seems to be the
soil relative humidity. It remains somewhat stable (over 80%) almost
throughout the wet period decreasing gradually afterwards and becomes

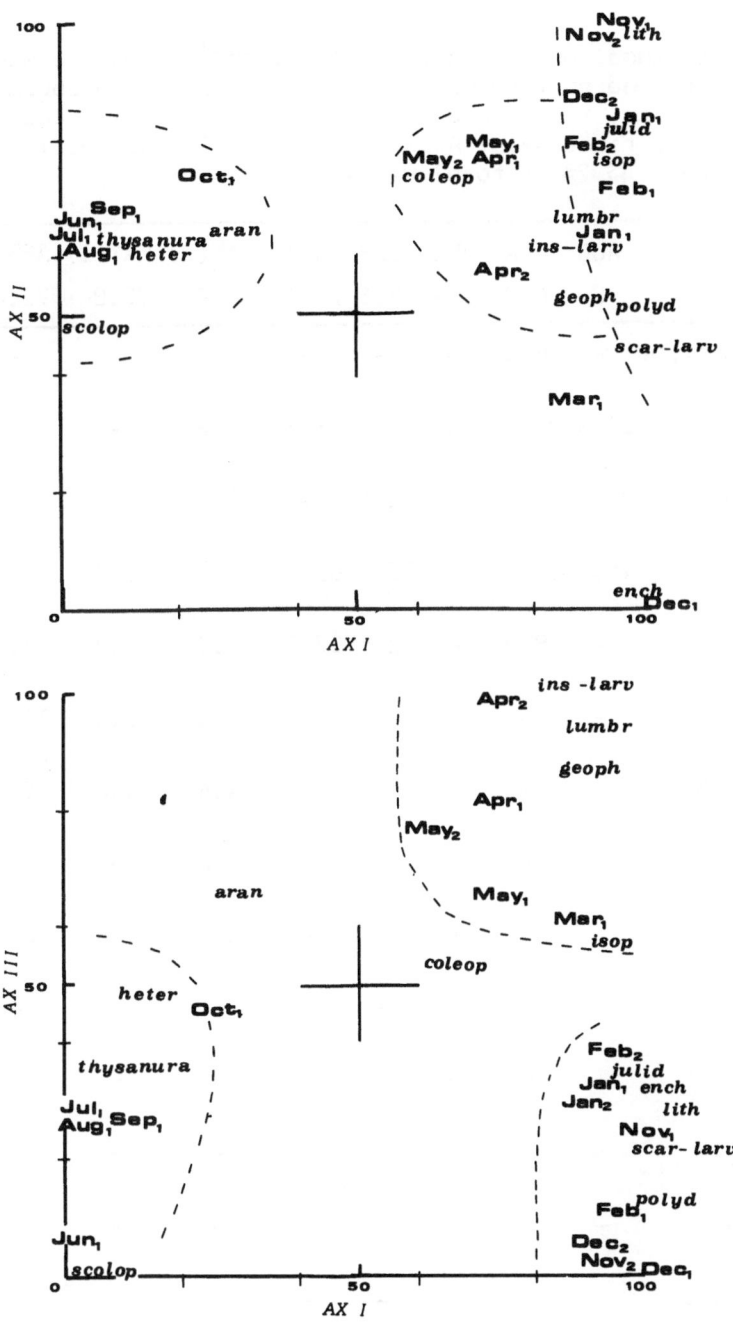

Fig. 1. R.A. 3 axes plane for macrofauna. The axes account 45.2, 16.8 and 13.8% of total variability, respectively. Months' indices correspond to the 1st and 2nd year measurements respectively.

Table 1. Macrofaunal densities (ind·m^{-2}·month^{-1}). For the summer period and corresponding groups only relative abudances are given. Months and animal groups are ranked according to R.A. first axis. Axis scores are also given. The axis account 45.2% of total variation.

	AUG	JUN	JUL	SEP	OCT	MAY(2)	MAY(1)	APR(2)	APR(1)
	0	0.2	0.6	8.5	23.1	60.6	72.0	72.4	73.1
Scolopendromorpha 0	-	6.2	-	-	-	-	-	-	-
Thysanura 1.6	25.8	43.7	27.3	25.0	9.1	1.6	-	-	1.6
Heteroptera 9.1	29.0	9.3	-	12.5	-	1.6	6.4	-	1.6
Araneae 25.3	45.1	28.1	45.5	37.5	63.6	8.0	5.6	14.4	9.6
Coleoptera 58.9	-	9.3	-	12.5	27.3	5.6	2.4	1.6	4.0
Insect Larvae 80.4	-	-	-	-	-	10.4	13.6	14.4	21.6
Geophilomorpha 84.1	-	-	-	-	-	1.6	1.6	2.4	4.0
Lumbricidae 84.7	-	-	-	-	-	5.6	-	20.0	6.4
Isopoda 90.9	-	-	-	-	-	1.6	1.6	2.4	-
Julidae 92.3	-	-	-	-	-	-	-	-	-
Scarab. Larvae 95.6	-	-	-	-	-	-	-	-	-
Enchytraeidae 96.0	-	-	-	-	-	-	-	5.6	-
Polydesmidae 96.1	-	-	-	-	-	-	-	-	-
Lithobiomorpha 100	-	-	-	-	-	-	-	-	-
Σx_t	?	?	?	?	?	36.0	31.2	60.8	48.8

| MAR(1) | JAN(2) | NOV(2) | DEC(2) | FEB(2) | JAN(1) | FEB(1) | NOV(1) | DEC(1) | X_t |
87.1	88.1	89.1	89.5	90.1	94.1	94.5	97.8	100	
-	-	-	-	-	-	-	-	1.6	?
-	2.4	1.6	4.0	-	-	-	-	-	?
2.4	-	-	2.4	1.6	-	-	-	-	?
6.4	6.4	4.0	17.6	8.0	4.0	1.6	-	4.0	?
4.0	2.4	30.4	16.0	6.4	2.4	4.0	2.4	4.0	?
28.0	24.0	7.2	9.6	13.6	14.4	5.6	5.6	12.0	10.00
8.0	6.4	1.6	5.6	4.0	2.4	1.6	-	5.6	2.50
8.0	24.0	-	8.0	21.6	17.6	-	18.4	9.6	7.70
2.4	-	-	6.4	8.0	12.0	5.6	7.2	1.6	2.70
12.0	80.0	110.4	164.0	58.4	58.4	25.6	59.2	50.4	34.35
2.4	8.0	3.2	5.6	2.4	1.6	2.4	-	9.6	1.95
26.4	33.6	-	12.0	8.0	5.6	31.2	-	129.6	14.00
-	-	-	10.4	4.0	1.6	-	-	6.4	1.24
-	-	-	2.4	-	1.6	-	4.8	1.6	0.57
100	187.2	158.4	264.0	136.0	121.6	77.6	97.6	236.0	

almost zero after May, following the rainfall decrease, and soil
evaporation increase due to high temperatures (Fig. 2a). Of course
differences are expected from year to year concerning the beginning
and the end of each period, because the precipitation pattern is
generally very variable (Fig. 2a).

More detailed analysis on the level of each distinguished period
is out of the scope of this work, although a partial disjunction
(transition from period II to III) may somewhat distort the pattern
(Gauch et al., 1977). The abrupt transition mentioned above is due
to the fact that the two main strategies, the soil animals show in
response to drought, avoidance and tolerance (Wallwork, 1976),
operate mainly at the end of period II. For example by the end of
period II almost all insect larvae have already become adult insects
(Fig. 2b) many of which leave the soil and consist a main fraction of
the drought resistant animals. This mechanism as well as the case
of migrations, which is the characteristic strategy for a variety
of soil animals, like Julidae in phrygana, are classified by Di
Castri in the category of active avoidance (Di Castri and Vitali-
Di Castri, 1981).

The group of Julidae coincide with the rainy season (Fig. 2b).
Though found into soil capsules in a curled position at about 30 cm
depth, averaging 3.3 individuals.m^{-2}, after early April they seem
to disappear from the upper soil layers. Under this inactive state
they pass all the drought period.

The trigger factor for Julidae vertical migration and inactivation at about the same time of the year (at least during the two
years we have been observing them) is not clear. This point according to our opinion needs careful study to be elucidated.

A detailed discussion on migrations of diplopods and the probable factors responsible for it are provided by Maija Peitsalmi
(1981).

Mesofauna

Acari (and among them Oribatei) and Collembola are the dominant
groups. The ratio of Acari to Collembola is much lower compared to
the ratios found in the Mediterranean formations of Chile (Di Castri
and Vitali-Di Castri, 1981).

The pattern of mesofaunal responses to time, analyzed by R.A.
after data transformation according to Wisconsin double standariz-
ation (Bray and Curtis, 1957), appears much more complex (Table 2).
Although the drought period can be distinguished here, as well, the
factor which governs the final result in the first axis is not so
simple as in the Macrofauna case. This is mainly due to rapid
responses of soil mesofauna to minor changes of moisture.

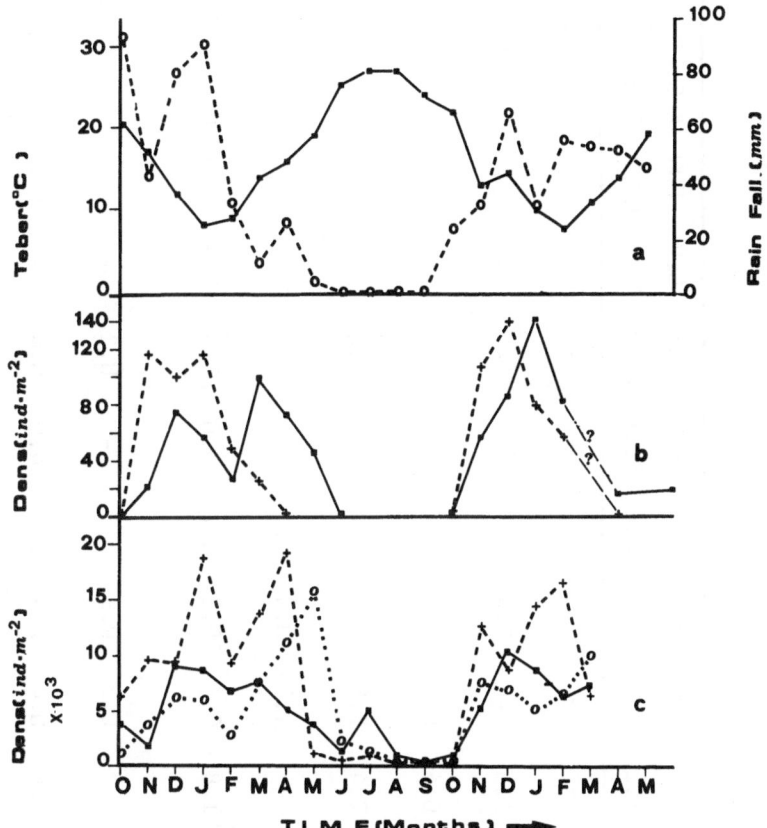

Fig. 2. a: Mean monthly temperatures (°C) •——• and precipitation
(mm), o‑ ‑ ‑o during the study period.
b: Julidae +‑‑ ‑‑+ and Insect larvae •——• densities
c: Mature Oribatei •——• , Juvenile o·····o and Collembola
+‑‑ ‑‑+ densities.

So, the position of July is indicative of the effects of an
accidental summer rain. A similar case appears after artificial
watering of sampling areas in September (SEP*).

According to the above, most of Mesofauna groups and especially
Acari appears to be active the year round, while some of them have
opportunistic strategies in response to drought. Mature Oribatei,
as a group, seem to have greater densities at winter gradually de‑
creasing during spring while their juveniles show a single peak at
May of the first year (Fig. 2c). Most of these juveniles belong
to a *Scheloribates* sp. Its mature individuals phenomenically show
low densities at summer but it is an artifact, as one can see in
Table 1 (densities of JUL and SEP*).

Table 2. Mesofaunal monthly densities (ind·11.63 cm^{-2}·month^{-1}). Months and animal groups are ranked according to R.A. first axis. Axis scores are also given. This axis accounts 25.7% of total variation.

	FEB 0.0	APR 2.8	JAN 4.3	DEC 6.6	MAR 7.1	JUL 11.8	NOV 30.6	MAY 32.0	SEP* 32.1	OCT 44.0	SEP 76.9	JUN 89.7	AUG 100	X_i
Polyxenidae 0	0.2	0.2	0.2	–	–	–	–	–	–	–	–	–	–	0.04
Symphyla 5.7	0.8	0.8	0.8	0.2	1.8	–	0.4	–	–	–	–	–	–	0.51
Protura 13.8	1	0.8	0.2	1.2	0.6	0.6	0.8	–	–	0.6	–	–	–	0.44
Oppiidae 13.7	15.6	2.8	8.8	23.4	7.4	2.8	0.6	3.4	5.5	4.4	–	0.4	–	5.77
Camisiidae 14.7	0.2	–	0.4	–	–	1.2	0.6	–	–	–	–	–	–	0.16
Collembola Arthropleona 14.9	37.2	70.6	74.2	24.0	51.8	2.6	35.2	3.6	3.0	23.4	0.4	2.0	0.6	25.27
Liodidae 18.0	–	–	–	0.2	–	0.2	0.8	–	–	0.2	–	–	–	0.09
Collembola Symphypleona 22.1	0.6	6.2	0.4	2.4	2.6	0.4	2.4	0.2	4.0	0.6	–	0.2	–	1.54
Isopoda 22.9	1.2	–	1.6	–	0.6	–	1.2	2.4	–	0.4	–	–	–	0.57

Hypochthonoidea 26.3	3.6	0.8	1.0	2.4	3.4	2.2	–	2.1	–	0.6	–	1.0	–	1.31
Scheloribates 27.9	4.4	17.2	18.6	4.8	15.4	21.8	1.6	6.6	11.0	2.6	0.4	3.4	1.8	8.43
Belboidea 31.2	0.2	0.2	0.6	0.4	0.4	0.4	0.2	0.4	–	0.2	–	–	0.2	0.24
Oribatei Juvenile 32.9	11.8	44.6	24.6	25.6	30.8	4.8	14.0	52.8	13.0	6.6	2.7	8.0	1.2	18.50
Acari exp. Oribatei 34.4	23.8	41.0	37.0	21.4	32.0	37.2	15.8	38.0	17.5	12.4	2.4	11.4	6.6	22.80
Insect Larvae 35.1	4.2	0.8	3.4	1.0	1.2	–	1.2	0.6	1.0	2.0	0.4	0.2	0.2	1.24
Pseudoscorpions 44.5	0.8	0.2	0.6	1.2	–	0.2	0.4	0.4	–	0.6	0.2	–	0.2	0.37
Coleoptera 55.2	–	1.0	–	–	1.0	0.2	–	0.6	0.5	1.8	0.2	0.4	0.2	0.45
Liacaroidea 69.8	–	–	–	–	–	–	1.0	0.2	–	0.2	–	–	0.4	0.14
Procidae 79.1	–	–	–	–	–	0.2	–	0.6	0.5	1.2	0.4	–	1.0	0.30
Thysanoptera 100	–	–	–	–	–	–	0.2	–	–	–	–	0.4	0.2	0.04
Σx_{ζ}	105.6	187.2	172.4	108.2	149.0	74.8	76.4	111.9	55.0	57.8	7.1	27.4	12.6	88.21

Vertical migrations in deeper soil layers controlled by moisture conditions may explain such a behavior (Usher, 1975; Metz, 1971; McBrayer et al., 1977).

As far as Collembola are concerned their densities are high from October to April, drastically decreasing afterwards. It seems that most adults die because of drought, after having left a certain number of resistant eggs. Similar conclusions are provided by Poinsot-Balaguer (1976) working in garrigue formations of South France. Comparison of the data on Arthropleona of JUN and SEP to those of JUL and SEP* (Table 2) does not provide any indication that species with aestivation mechanisms, as the previous author reports (Anhydrobiosis), appear in phrygana. The mechanism by which some species of low densities seem to remain active throughout the dry period (Table 2) cannot be yet explained.

On the other hand, it must be pointed that a species of the family Sminthuridae, shows a relatively high density at SEP*, fact which may be due to a rapid development from eggs. Wetting dry soil, in preliminary experiments, we observed that this species could develop in only four days.

Decomposition processes in phrygana

Soil animals are generally considered to have a synergetic role in decomposition with decomposers. Some of the ways by which soil fauna influences decomposition rates are discussed recently by Luxton (1982),and Cansela da Fonseca and Poinsot-Balaguer (1982).

Decomposition rates during a two years period are provided by Fousseki and Margaris (1981). Some of those appear in Table 3. The relative growth rate of producers are taken from Margaris (1976). Again three periods are considered as in the macrofauna case. Each value is the average of mean monthly estimations of each parameter. A seasonal pattern is observed, very similar to that of the animal activities. This synchronization seems to be under climatic determination. Nevertheless, a possible role of soil macrofauna cannot be excluded.

The first autumn rains is the stimulus for producers and de-composers activity. During autumn maximal activity of macrofauna is observed (Table 1) resulting to the development of a separate layer (3 cm deep) of detritus and excrements. Higher concentration of ash elements as compared to the litter consumed are expected in this layer (Striganova and Valiachmedov, 1976; Koslovskaja and Striganova, 1977; Richter, 1979). This layer consists the substrate of microbial growth, very high during that period as measured by dehydrogenase and nitrifying activity (Fousseki and Margaris, 1981). Bacterial colonization of animal faeces is possibly favoured by the compact form of finely comminuted plant material in them

Table 3. Some data on climate, decomposition* and producers R.G.R.**
according to the 3-period division of the year.

	Per. I	Per. II	Per. III
Mean Temperature of the Period	11.67 ± 2.18	15.65 ± 2.68	24.92 ± 1.47
Mean Minimum Temperature	8.42 ± 2.68	10.30 ± 3.52	20.39 ± 1.76
Mean Maximum Temperature	15.24 ± 2.62	18.29 ± 4.15	22.02 ± 2.00
Mean Rainfall	55.33 ± 16.14	32.28 ± 17.7	7.24 ± 56.66
Maximum Rainfall	91.2 (JAN 81)	54.4 (MAR 82)	23.2 (OCT 81)
Minimum Rainfall	11.7 (NOV 81)	3.6 (MAY 81)	0 (all summer)
Mean Dehydrogenase Activity $ng\ TPF.g^{-1}$	0.237 ± 0.06	0.103 ± 0.06	0.095 ± 0.06
Nitrifying Activity $g\ NO^{-3}.g^{-1}.day^{-1}$	2.616 ± 0.22	1.95 ± 0.83	0.95 ± 0.45
NO^{-3} ions $\mu g.g^{-1}$	25.3 ± 5.19	17.66 ± 1.99	22.46 ± 3.65
Plant R.G.R.	3.62 ± 1.07	0.66 ± 1.94	-3.05 ± 4.4

*Fousseki and Margaris (1980)
**Margaris (1976)

(Hanlon, 1981a, 1981b). Furthermore, high mesofaunal activity has
been proved to occur in the same layer (Sgardelis, unpublished data).

The increased macrofaunal activity inducing activation of the
whole decomposition process results to high nutirents availability
right at the favourable time for plant growth when their needs in
nutrients are maximized.

The impacts of macrofauna determining the decomposition process
is extended even after macrofauna inactivation by offering an intermediate material for the decomposers complex to go on.

Concluding, it can be said that phryganic ecosystems are characterized by intense seasonality with short favourable periods in
which all procedures are synchronized. We believe that the synchronization feature is extremely interesting and must constitute a subject
of further detailed studies.

REFERENCES

Bray, J. R., and Curtis, J. T., 1957, An ordination of the upland
    forest communities of Southern Wisconsin, Ecol. Monogr., 27:
    325-349.
Cancela da Fonseca, J. P., and Poinsot-Balaguer, N., 1982, Regime
    alimentaire des microarthropodes du sol en relation avec la
    decomposition de la matière organique, C. R. Acad. Paris (in
    press).
Di Castri, F., and Vitali-Di Castri, V., 1981, Soil Fauna of Medi-
    terranean-climate regions, in: "Ecosystems of the World. II.
    Mediterranean-type Shrublands," Di Castri, Goodall and Specht
    eds., Elsevier Scientific Publishing Co., Amsterdam, Oxford,
    New York.
Fousseki, E., and Margaris, N. S., 1981, Soil metabolism and de-
    composition in a phryganic (East Mediterranean) ecosystem,
    Oecologia, 50:417-421.
Gauch, H. G., Jr., Whittaker, R. H., and Wentworth, T. R., 1977,
    A comparative study of Reciprocal Averaging and other ordination
    techniques, J. Ecol., 65:157-174.
Hanlon, R. D. G., 1981a, Some factors influencing microbial growth
    on soil animal faeces. I. Bacterial and fungal growth on
    particulate oak leaf litter, Pedobiologia, 21:257-263.
Hanlon, R. D. G., 1981b, Some factors influencing microbial growth
    on soil animal faeces. II. Bacterial and fungal growth on
    soil animal faeces, Pedobiologia, 21:264-270.
Hill, M. O., 1977, Reciprocal averaging: an eigenvector method of
    ordination, J. Ecol., 61:237-249.
Kozlovskaja, L. S., and Striganova, B. R., 1977, Food, digestion
    and assimilation in desert woodlice and their relations to
    the soil microflora, in: "Soil Organisms as Components of

Ecosystems," U. Lohm and T. Persson, eds., Ecol. Bull., Stockholm.

Luxton, M., 1982, General ecological influence of the soil fauna on decomposition and nutirent circulation, Oikos, 39(3):355-357.

Margaris, N. S., 1976, Structure and dynamics in a phryganic (East Mediterranean) ecosystem, J. Biometeor., 3:249-259.

Margaris, N. S., and Vokou, D., 1982, Structural and physiological features of woody plants in phryganic ecosystems related to adaptive mechanisms, Ecologia Mediterranea, T.VIII, Fasc 1/2: 449-459.

McBrayer, J. F., Ferris, J. M., Metz, L. J., Gist, C. S., Cornaby, B. W., Kitazawa, Y., Kitazawa, T., Wernz, J. G., Krantz, G. W., and Jensen, H., 1977, Decomposer invertebrate populations in U.S. forest biomes, Pedobiologia, 17:89-96.

Metz, L. J., 1971, Vertical movement of Acarina under moisture gradients, Pedobiologia, 11:262-268.

Nahal, I., 1981, The Mediterranean climate from a biological view-point, in: "Ecosystems of the World. 11. Mediterranean-type Shrublands," Di Castri, Goodall and Specht, eds., Elsevier Scientific Publishing Co., Amsterdam, Oxford, New York.

Peitsalmi, M., 1981, Population structure and seasonal changes in activity of Proteroiulus fuscus (Am Stein) (Diplopoda, Blani-ulidae), Acta Zoologica Fennica, 161:1-66.

Poinsot-Balaguer, N., 1976, Dynamique des communantes de Colles en milien xerique mediterraneen, Pedobiologia, 16:1-17.

Richter, K. O., 1979, Aspects of nutrient cycling by Ariolimax columbianum (Mollusca: Arionidae) in Pacific Northwest coni-ferous forests, Pedobiologia, 19:60-74.

Schaefer, R., 1973, Microbial activity under seasonal conditions of drought in Mediterranean climates, in: "Mediterranean-type Ecosystems. Origin and Structure," F. Di Castri and H. A. Mooney, eds., Springer-Verlag,Berlin, Heidelberg, New York.

Striganova, B. R., and Valiaschmedov, B. V., 1976, Beteiligang boden-bewohnender Saprophagen an der Zersetzung der Laubstreu in Pistazienwäldern, Pedobiologia, 16:219-227.

Usher, M. B., 1975, Seasonal and vertical distribution of a population of soil arthropods: Cryptostigmata, Pedobiologia, 15:364-374.

Wallwork, J. A., ed., 1976, "The Distribution and Diversity of Soil Fauna," Academic Press, London.

COMPARATIVE STUDIES OF ORTHOPTERAN SPECIES ADAPTED TO LIVING ON THE GROUND AND OF SOME STRONG FLIERS FROM THE SAME ORDER

G. Theophilidis

Laboratory of Animal Physiology
School of Physics and Mathematics
University of Thessaloniki, Greece

INTRODUCTION

Field studies of the behavior of some orthopteran species, found in wheatfields and the bush surrounding them, show that they can be classified according to their behavior into three categories a) insects adapted to live actively on the ground; these are excellent runners, b) species which are able to produce a powerful jump; these are neither good in walking nor in flying and c) insects which are strong fliers (very few species belonging to Acrididae). Since in all these behavior patterns the metathoracic legs and especially the metathoracic extensor tibiae muscle (meta ETi), the jumping muscle, plays an important role, the anatomy and innervation of this muscle was examined in the three categories described above. It seems that in each case the metaETi and also some other morphological and anatomical features of the examined species have been adapted to meet the requirements of their behavior and their environment.

MATERIALS AND METHODS

Field studies took place in an area of about 1,500 square meters of wheat-fields surrounded by thick green bush, 20 km east of the city of Thessaloniki, during the period of summer and the beginning of the autumn (July to the end of October). The most dominant species were *Decticus Albifrons* (Tettigoniidae), *Calliptamus* sp. (Catantopidae) and *Acrida mediterranea* (Acrididae). The first two were mainly found during July and August while the last one during September and October.

Table 1. The values represent the mean effective diameters from four different insects of the same species. The standard deviations were between 1.20 to $\pm$ 3.5 μm for the FETi, SETi and inhibitor (I). For the DUM neurons was between $\pm$ 0.3 to $\pm$ 1.2 μm. + The values for the diameters of the locust ETi axons were obtained from Hoyle (1978). * The larger from the DUM neurons. T.l/F.l: Tibia to femur length, A.l/b.l: Antenna to body length, B.w/B.l: Body weight (dry weight) to body length, FETi: Fast Extensor Tibia, SETi: Slow Extensor Tibia, DUM: Dorsal Unpaired Median (Hoyle, 1974).

	FETi	SETi	I	DUM	FETi/SETi	T.l/F.l	A.l/B.l	B.w/B.l
Decticus Ablifrons								
Meta.v.ex.n.		17.5	8.3	2.5*		0.95	1.40	0.30
Calliptamus sp.								
Meta.v.ex.n.	23.0	13.0	7.4	3.5	1.76	0.78	0.40	0.10
Acrida mediterranea								
Meta.ex.n. (proximal)	18.0	12.0	6.0	3.0*	1.50	0.96	0.27	0.05
Meta.ex.n. (distal)	10.1	8.2	4.1	2.0				
Schistocerca gregaria[+]								
Meta.ex.n.	18.0	14.0	6.0	3.0	1.28			

To study the anatomy of the metaETi, the femur was removed and fixed (ventral side up) on a plasticine platform. To expose the ETi muscle and its motor nerve the ventral surface of the femur and the flexor tibiae muscle was removed. The extensor nerve was back-filled through N5 (nomenclature form Campbell, 1961) with $CoCl_2$ or $NiCl_2$. Methylene blue was also used to establish the details of the innervation of the metaETi.

For electron microscopy studies, the metaETi and its motor nerve were fixed, (Karnovsky, 1965) postfixed in OsO_4 1%, dehydrated, and mounted in a low viscosity epoxy. Thin transverse sections (70 nm) and thick sections (1-5 μm) were cut from the extensor nerve at different levels using a Reichert Ultratome. Light straw to silver-grey sections were mounted on 200 mesh grids and double stained in lead citrate and uranyl acetate. All grids were examined in a Jeol JEM 100B electron microscope at 80 KV.

For electrophysiological studies, the animals were mounted on a plasticine platform (ventral side up),and the femurs of the pro-meso- and metathoracic legs were also fixed on the platform. The ETi was exposed, as described above. The use of saline was avoided but when it was required, locust saline (Usherwood and Grundfest, 1965) was used. The recordings from the extensor nerve were made *en passant* using a hook electrode (modified by Wilkens and Wolf, 1974) or a gold-plated suction electrode (Theophilidis and Burns, 1982). The nerve impulses were amplified using a Phipps and Bird Bioamplifier and displayed on a Tektronix oscilloscope.

RESULTS

Decticus Albifrons

This is a relatively large insect with a body length of about 5 to 6 cm (Fig. 1A). It lives in the lower parts of thick bush where it can move and climb very fast. Sometimes it can be seen in open ground but as soon as it is aware of danger it runs extremely fast to hide in the bushes or under stones, if they are available. Occasionally, it jumps and flies 5 to 10 meters but only when it is in open ground. It has a heavier body than the other species examined with a ratio of body weight to body length of 0.3. The same ratio for the other species is smaller (Table 1). Also it has relatively long antennae, sometimes longer than its body length while in all the other species they are shorter (Table 1 and Fig. 1). The antennae are thin, flexible and very well controlled, capable of moving very fast in all directions.

The Metathoracic ETi

Although the pro- and mesothoracic femurs of *Decticus Albifrons*

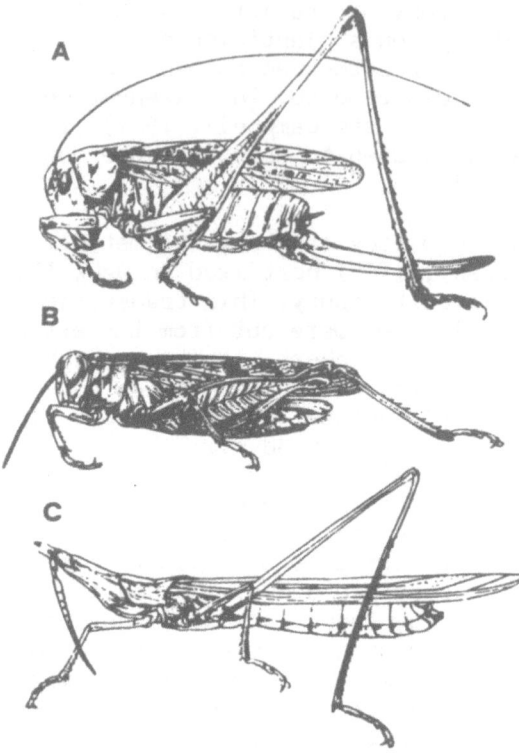

Fig. 1. A. *Decticus Albifrons* (X 1.3), B. *Calliptamus italicus* (X 2)
and C. *Acrida mediterranea* (X 1). (The drawings were obtained
from L. Chopard, 1951).

are almost identical to the homologous legs of the locust, the
metathoracic legs are different in shape. The first half of the
femur has semispherical shape (Fig. 2A) while the rest suddenly
becomes very narrow, with a cylindrical shape (external diameter
of 1.5 to 2 mm). The metathoracic ETi is concentrated only in the
first half of the femur. It is pinnate in form and it consists of
14 to 17 pairs of muscle bundles. They are very long (approx. 5 mm)
and unusually large in the proximal part of the femur with a
pinnation angle of 15°. In the distal part of the ETi (almost in the
middle of the femur) they become very small (approx. 1 mm long)
while their pinnation angle increased to 30°. Only the apodemes of
the flexor and the extensor muscle were found in the cylindrical
part of the femur. The main femoral proprioceptor, the Chordotonal
organ (CO), was found in the proximal part of the femur while it is
always in the distal part, in locust (Usherwood et al., 1968)
Calliptamus sp. and *Acrida mediterranea*.

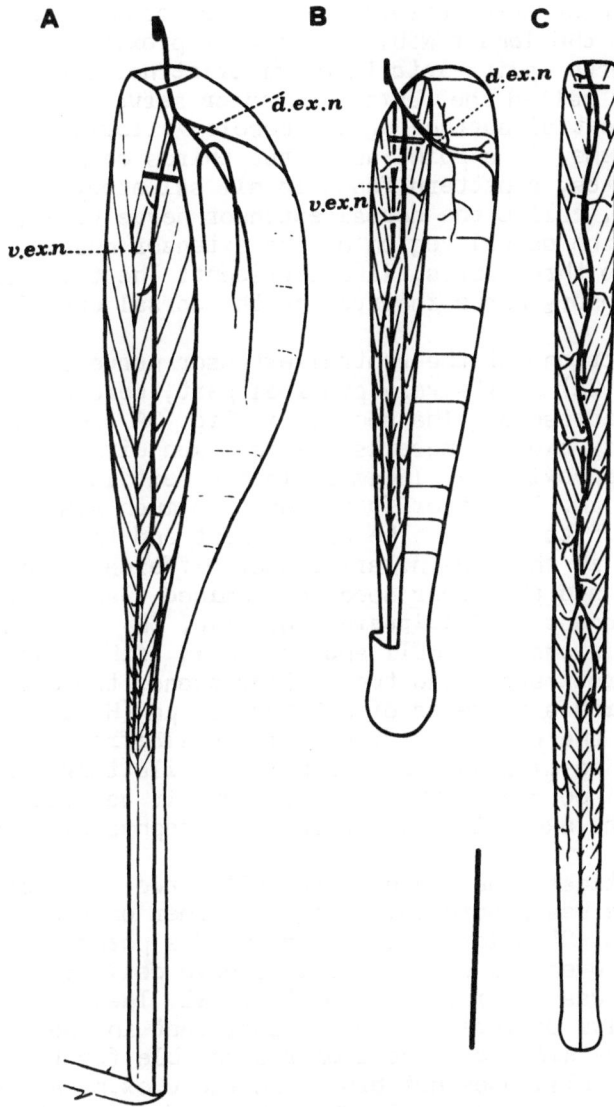

Fig. 2. Drawings of dissected femurs from various orthopteran
species showing the motor nerve, supplying the extensor
tibiae (ETi), and its branching pattern. The flexor tibiae
and retractor unguis muscles as well as the trachea were
dissected away. A. Metathoracic femur of *Decticus Albifrons*.
B. *Calliptamus* sp. and C. *Acrida mediterranea*. The vertical
lines on the proximal part of the extensor nerve indicate
the area of the motor nerve where the cross-section for
the Electron Microscopy were taken. v.ex.n.:ventral extensor
nerve, d.ex.n.:dorsal extensor nerve. Scale bar: 6 mm.

The extensor nerve enters the muscle dorsally through the equivalent to the locust N5B. At the very proximal part of the muscle the nerve splits into three district nerve branches. The first branch, called the ventral extensor nerve (v.ex.n., Fig. 2A), lies on the ventral surface of the muscle following the line of the extensor apodeme. At a point about two thirds down the muscle, the nerve bifurcates, a pattern found in all saltatory species. The second branch, called the dorsal extensor nerve (d.ex.n., Fig. 2A) dives inside the dorsal region of the extensor muscle bundle where it splits into three parts, like a trident. The third part is probably a sensory nerve innervating the dorsal cuticle of the femur.

Cross sections of the ventral extensor nerve (the two vertical lines in Fig. 2A) at its very proximal part, were examined using an electron microscope. The section in Fig. 3A shows the profiles of at least six axons. There is a large axon with an effective diameter of approximately 16 µm while the next largest axon has a diameter of 7 µm. The other four axons are much smaller with a diameter of 1.5 to 0.4 µm. The arrangement and the dimension of the axon profiles in the v.ex.n. are rather different from those found in the locust and the other species examined (Table 1 and Fig. 3B, 3C). In the locust of *Calliptamus* sp. (Fig. 3B) for example, there are two large axons, the FETi and the SETi, with a diameter of 18 µm and 14 µm approximately and two smaller axons, the common inhibitor and DUMETi, with a diameter of 6.0 and 3.0 µm (Hoyle, 1978). It seems that in the v.ex.n. of the *Decticus Albifrons* one of the largest axons is missing (FETi or SETi), while there are some extra smaller axons (possibly DUM neurons). Thus it was found necessary to identify the axons in the v.ex.n. electrophysiologically.

To investigate the route of the FETi, two stimulating electrodes were used. One was placed on N5 and the other on the very proximal part of v.ex.n. Stimulation of N5 produced a powerful fast tibia movement, but when the electric shocks were delivered only on the v.ex.n. there was no tibiae movement at all. These experiments, repeated on five different animals, show that in the metathoracic leg of *Decticus Albifrons* the axon responsible for the fast tibia movement, the FETi, does not branch in the v.ex.n. but innervates the extensor tibia muscle only through the dorsal extensor nerve (d.ex.n., Fig. 2A).

The fact that stimulation of the v.ex.n. does not produce any tibiae movement suggests that the two large profiles in Fig. 3A (20 µm and 8 µm in diameter) belong to the SETi and the inhibitor neurons. Records from the v.ex.n. of an intact insect reveal the spikes of only two large axons (Fig. 4A). The axon producing the largest spike (SETi Fig. 4A) is more active and responds more easily to any tibiae flexion, a resistance reflex typical of the SETi neuron. The small spike belongs to a less active neuron which always responds to tibiae extension (I, Fig. 4A), a resistance

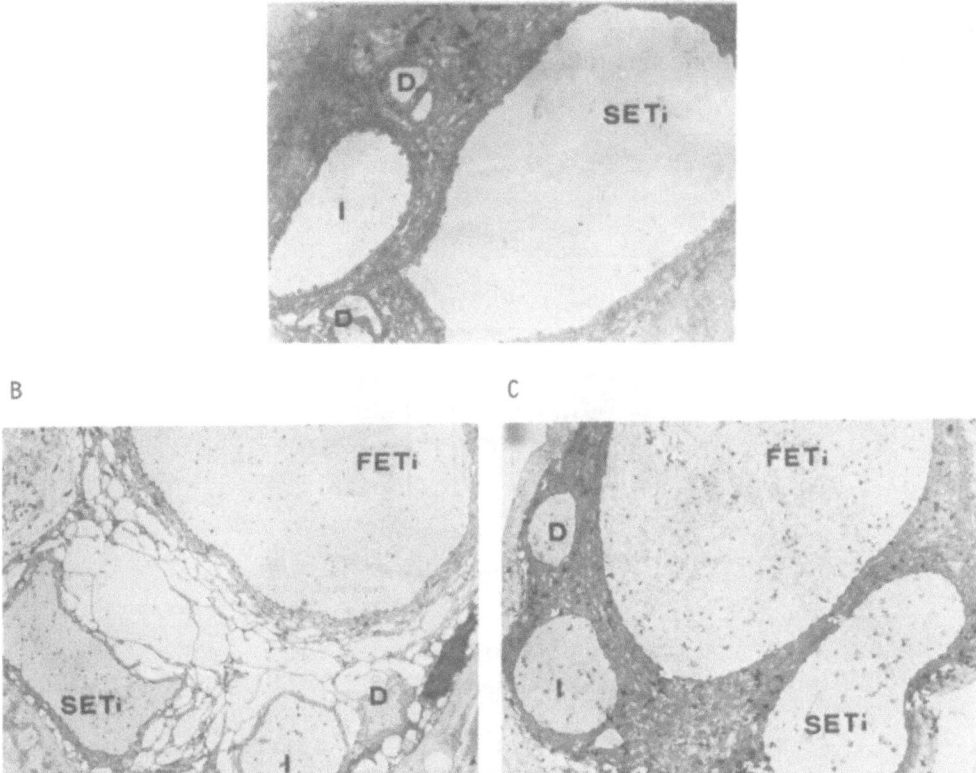

Fig. 3. Electron micrographs of transverse sections from the proximal
        part of extensor motor nerve of three different orthopteran
        species. A. The metathoracic v.ex.n. of *Decticus Albifrons*.
        B. The metathoracic v.ex.n. of *Calliptamus* sp. and C. The
        metathoracic ex.n. of *Acrida mediterranea*. From the exact
        position of the sectioning area see vertical lines on the
        extensor nerve in Fig. 2.
        A,C X 3000, B X 2.200.
        SETi: Slow Extensor Tibia, FETi: Fast Extensor Tibia,
        I: Inhibitor, D: DUM, Dorsal Unpaired Median axon.

reflex typical of an inhibitory neuron.

        At least three of the very small axons (Fig. 3A) belong to DUM
neurons. This was demonstrated by stimulating the v.ex.n. and
recording from its contralateral nerve (Hoyle et al., 1974; Hoyle,
1974). When the intensity of the stimulus applied to the v.ex.n.
was gradually increased three spikes D1, D2 and D3 (Fig. 4B) were
recorded in the contralateral nerve indicating the existence of

three DUM neurons with different thresholds. These experiments were repeated on six different animals and three DUM neurons were always found innervating not only the metathoracic but also the pro- and mesothoracic extensor nerves.

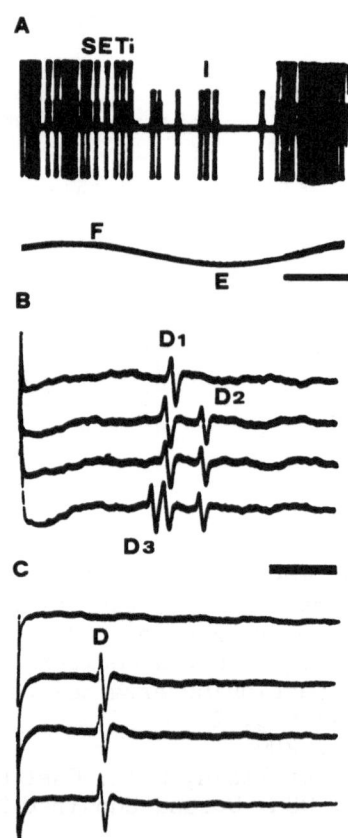

Fig. 4. A. Extracellular records from the metathoracic ventral
extensor nerve (v.ex.n. see Fig. 1A) to demonstrate the
activity of the SETi and inhibitor (I). The second trace
represents tibia movement. B,C. As in A, while intensity
of the stimulating electrode on the contralateral v.ex.n.
was gradually increased. (B in *Decticus Albifrons*, C in
Calliptamus sp.). D_1,D_2,D_3,D:DUM neurons. E:tibial extension.
F:tibial flexion. Calibration: A 20 ms, B, C 10 ms.

The other axons from the metathoracic FETi and inhibitor neurons enter the muscle through N5 and the axon from SETi through N3B (Fig. 5A).

Calliptamus sp.

This is a relatively small insect with a length of 28 to 30 mm (Fig. 1B). The body weight to body length ratio is approximately 0.1, a value found for at least four other similar species examined. It was mainly on the top or upper parts of grass or wheat. For fast transportation (escape behavior, for example) it produces a powerful jump which is usually completed by flying or hovering for 2 or 3 meters. It is not a strong flier, compared to the locust. For posture, climbing and walking it uses mainly the pro- and mesothoracic legs but it can never walk or run long distances as Decticus Albifrons does.

The metathoracic femur is about 20 mm long, the mesothoracic 5 mm and the prothoracic 3 mm. In Calliptamus sp. there is a difference in the construction of the metathoracic leg. The tibia is considerably shorter than the femur with a ratio of tibia to femur length of 0.78 while the same ratio for all the other species is approximately 1.0 (Table 1). The significance of the shorter tibia in the jumping activity of the insect will be discussed later in the text. The metathoracic ETi muscle has a construction similar to the homologous muscle of the locust except that no accessory extensor muscle bundles were found in Calliptamus sp. (Fig. 2B). There are about 17 muscle bundles. At the proximal part of the femur they are 3 or 4 mm long and have a pinnation angle of 25°. At the distal part they become smaller (approximately 1.5 mm) and the pinnation angle increases up to 45°. The extensor nerve enters the muscle dorsally through N5B1 as it does in the locust but at the proximal part of the muscle it splits into three distinct nerve branches: the ventral extensor nerve, the dorsal extensor nerve and a long sensory nerve (Fig. 2B). The same pattern was found in Decticus Albifrons with some differences in the branching pattern of the d.ex.n. Electron Microscopy sections of the ventral extensor nerve show that it always contains four axons (Fig. 3B and Table 1). They were identified as the FETi, SETi, inhibitory and DUMETi. The DUMETi was identified by stimulating the v.ex.n. and recording from its contralateral (Fig. 4C). The FETi and the DUMETi enter the muscle through N5 while the other two (SETi, inhibitor) enter from N3 (Fig. 5B).

Acrida mediterranea

This is a large insect about 5 cm long (Fig. 1C). It is relatively light in weight having a ratio of body weight to body length of 0.05 (Table 1). It was found mainly on the upper parts of crops. It can jump and fly for 3 to 15 meters.

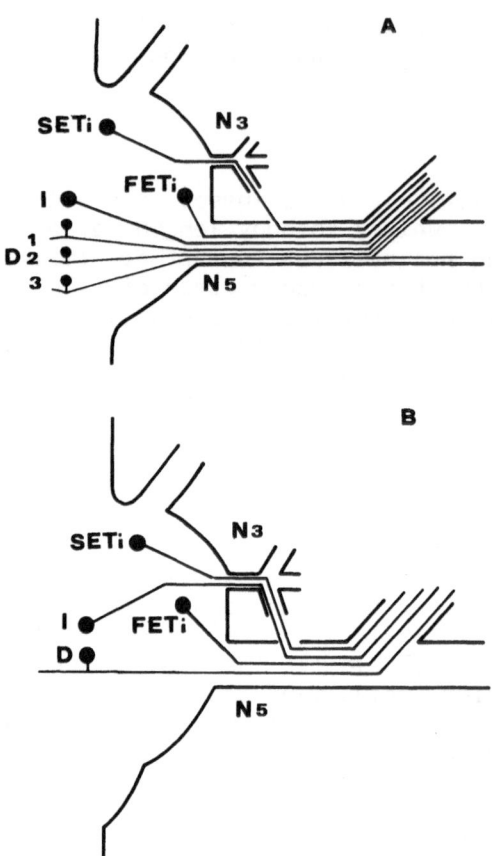

Fig. 5. A. The positions of the extensor tibiae motor axons in the
metathoracic nerves of *Decticus Albifrons*. B. The positions
of the extensor tibiae motor axons in the metathoracic nerve
of *Calliptamus* sp. and *Acrida mediterranea*. In none of the
three species the exact positions of the motor neurons are
known. The neurons are placed in the drawings according to
the positions of the homologous neurons of the locust.
(Burrows, 1973; Burrows and Hoyle, 1973; Hoyle et al., 1974).

The metathoracic femur is 30-32 mm long having a cylindrical
shape. It has an effective diameter of approximately 2 mm.. The
pro- and mesothoracic legs are about 5 and 7 mm long. They also have
a cylindrical shape but they are very thin, with an effective
diameter of 0.5 mm. The metathoracic ETi muscle occupies the dorsal
half of the femur and it has a pinnate form. It consists of 35 to

40 pairs of small muscle bundles which are 2.5 to 3.0 mm long in the proximal part of the muscle with a pinnation angle of 15°. The muscle bundles of the distal part of the muscle are small, about 1.5 mm to 2.0 mm long having a pinnation angle of 40°. The extensor nerve runs parallel to the apodeme of the muscle but it has a zig-zag shape (Fig. 2C). At a point about two thirds down the femur, a Y fork occurs. Electron Microscopy sections of the extensor nerve at its proximal end, show that it always contains five axons (Fig. 3C and Table 1). The three largest have been identified physiologically as the FETi, the SETi and the inhibitor in descending size order. The other two axons seem to come from the DUM neurons. However, no spikes were ever recorded in the extensor nerve when its controlateral nerve was stimulated. This suggests that in this insect the DUM neurons may have a different shape from the homologous neurons of the locust and *Calliptamus* sp. (Fig. 4B, 4C). Electron Microscopy sections of the extensor nerve, approximately 0.5 mm before the bifurcation point, show that one of the smallest axons does not continue up to that point. The axon FETi enters the muscle through N4 while the inhibitor and the SETi, enter from N3, the same pattern as in *Calliptamus* sp. and the locust (Fig. 5B).

DISCUSSION

Acrida mediterranea, Calliptamus sp.

There are strong similarities in the basic construction of the metathoracic ETi muscle between the two species and the locust. For example, the extensor nerve contains four axons, in all cases, the FETi, SETi, DUMETi and the inhibitor, found in almost every branch of the extensor nerve (common innervation, Hoyle, 1955) tapering, especially the FETi, towards the distal end of the muscle. In both species and the locust the FETi and the DUMETi (Hoyle, 1978) leave the metathoracic ganglion through N5, and the SETi and the inhibitor reach the ETi through N3. There is also a similarity in the axon diameters and although the two examined species and the locust have different body length and weight the effective diameter of their FETi is 18 to 24 μm their inhibitor is between 6 to 7 μm.

Other strong similarities are that the metathoracic ganglion in all cases is fused with the first abdominal and finally, although there are some differences in the basic construction of the meta-thoracic chordotonal organ in all species, it is located at the distal part of the femur as it is in the locust (Usherwood et al., 1968).

In spite of these similarities there are anatomical and morpho-logical differences which seem to be related to the behavioral differences found between the examined species. As an escape behavior, for example, *Acrida mediterranea*, as the locust, jumps to take off and continues to fly for 10 to 15 meters. In contrast, *Calliptamus* sp. and other similar species produce a powerful jump to gain height and then they fall to the ground hovering or they can fly for a few meters (2 or 3) before they land. For each behavior pattern, described above, there seem to be certain adaptations. For example, species which fly mainly (locust, *Acrida mediterranea*) are lighter in weight, have longer wings and in some cases as in the locust, a strong stream of air in the front part of the head initiates flying. Species which mainly jump also have some adaptations, especially in their meta-thoracic legs. In *Calliptamus* sp., for example, although the ETi has a similar shape and construction to the homologous muscle of the locust, the proximal dorsal part of the muscle is innervated by an extra nerve branch (d.ex.n.) containing also the FETi. This is probably a way of increasing the power output of the muscle, vital for long jumps, by increasing the number of muscle fibers which are activated by the FETi. The FETi is relatively larger from all the other species (23 μm FETi/SETi 1.76, Table 1) and probably this is another adaptation in order to increase the jumping performance of *Calliptamus* sp. Another adaptation which can be considered is the fact that in *Calliptamus* sp. the length of the tibia is shorter than its femur (ratio of tibia to femur length 0.78) while in most species adapted to flying they are equal. A shorter tibia means that the distance between the load at the end of the tibia and the femural tibiae pivot is decreased. This increases the mechanical advantage of the metathoracic ETi of *Decticus Albifrons* in relation to the other species where the tibiae is relatively longer.

Decticus Albifrons

Its legs have a construction similar to that of the other ortho-ptera. Its pro- and mesothoracic femural muscles are almost identical to those of the locust (Burns and Usherwood, 1979; Theophilidis and Burns, unpublished data), but its metathoracic ETi have the follow-ing basic differences from all the other species: first, the main mass of the muscle is suppressed and concentrated in the first half of the femur. There is a strong contrast in the arrangement of the muscle bundles which are large and well developed in the proximal part, while they become smaller and thiner almost in the middle of the femur. In all the other species the ETi occupies almost the full length of the muscle and there is not such a strong contrast in the muscle bundle arrangement, even in *Calliptamus* sp.

Second, the innervation of the ETi muscle is different from other orthoptera. *Decticus Albifrons* is the only reported case where the metathoracic ETi has separate innervation. In all the other species including the locust (Hoyle, 1955), the same muscle has a common innervation. The axon FETi, responsible for fast tibia move-ment, does not innervate the ventral fase and the distal part of the muscle since it was never found in cross-sections of the ventral bifurcating extensor nerve. It innervates only the proximal well-developed area of the muscle through the d.ex.n.

Third, the unusual shape of the femur seems to have some effect on its internal construction. The main femoral proprioceptor, the chordotonal organ (CO), is located at the very proximal dorsal anterior part of the femur, having an apodeme of 3 to 4 cm long (Theophilidis, unpublished data). In constrast, the homologous organ of the locust (Usherwood et al., 1968) and of all other species examined was found in the very distal part of the femur. A possible explanation for the unusual position of the CO could be the fact that the metathoracic femur becomes very narrow in the distal part, so the development of such a large sensory organ is prevented. Some other possible explanations will be discussed further in the text.

Finally, there are also some other differences which seem to be related to the internal organization of the ganglion. The ETi is innervated by two extra DUM neurons. The inhibitory neuron reaches the ETi muscle through N5 and not through N3 and the metathoracic ganglion of *Decticus Albifrons* is not fused with the first abdominal and it has the shape of the pro- and mesothoracic ganglion.

At this point innevitable arises the question: Why is the *Decticus Albifrons* so different from all the other orthoptera? A possible explanation for this difference in construction, could be the fact that while all the examined species have the tendency to avoid the ground by jumping, flying and living on the upper parts of bushes or crops, *Decticus Albifrons* has the tendency to live a very active life on the ground. Thus apart from the fact that it is a relatively heavier insect it has some other vital adaptations for its survival on the ground. It has much longer antennae which are thin, flexible and very well controlled, capable of moving very fast in all directions. Probably it uses the long antennae to detect the large number of obstacles found in the ground and to obtain a better three dimensional "view" of its immediate environment. This kind of sence is important since the lighting condition in the lower parts of thick bushes are very bad during the day and *Decticus Albifrons* has also an active nocturnal life. One of its best adaptations to such an environment is that it can walk and run extremely fast, a vital condition for survival on the ground. The ability of *Decticus Albifrons* to run extremely fast combined with the fact that it is a relatively heavier insect indicates that it uses its leg muscles to a greater extent than all the other species. Thus it is not surprising that its pro-, meso- and metathoracic ETi are supplied with two extra octopaminergic (DUM) neurons. The similar neuron innervating the jumping muscle of the locust, the DUMETi, was found to mediate: a) the slowing of a myogenic rhythm found in a specialized proximal bundle of muscle fibres, b) the increase in amplitude of slow motoneuron twitch tension and c) the increase in relaxation of twitch tension induced by firing either the fast or the slow motoneurons (Evans, 1981).

The walking pattern of *Decticus Albifrons* is not much different
from the locust pattern described by Burns and Usherwood (1979).
However, preliminary studies show that it can achieve a high running
speed by using its metathoracic ETi muscle. Activation of the FETi,
which in the locust is generally silent in fast walking, creates
quick short jumps which accelerate the insect. These jumps are
extremely fast, accurate and well coordinated with all the other
legs. This contrasts with most of the other orthoptera, in which
the metathoracic FETi is used mainly for the massive activation of
the ETi. Of course such an extensive use of the metathoracic ETi
in fast walking requires a better and more accurate neuronal control
of the jumping muscle. It seems that in *Decticus Albifrons* this is
achieved by the following modifications: a) by the transposition
of the femoral chordotonal organ towards the proximal part of the
femur. The fact that this sensory organ is located 3 to 3.5 cm
closer to the ganglion indicates that the hysterésis between the
activation of the sensory input and the arrival of the sensory information at the CNS is minimized. This is vital for the faster
response and the better control of the jumping muscle. Possibly
this is also one of the reasons why the CO in the pro- and mesothoracic legs which are used mainly for posture, and walking, is
also located in the proximal part of the femur (Burns, 1974) and
not at the distal and b) by changing the innervation of the jumping
muscle from a common to a separate pattern. The advantage of separate
innervation in the neuronal control of the muscle is that different
muscle regions can play different roles functionally. In *Decticus
Albifrons* this differentiation in the muscle regions is not limited
to the fact that they are differentially senstitive to the pattern
of the SETi discharge, as it was suggested for the locust by Hoyle
(1978), but it is emphasized by the fact that the distal region of
the muscle is innervated only by the SETi while the FETi was found
only in the proximal dorsal regions.

REFERENCES

Burns, M. D., 1974, Structure and physiology of the locust femoral
     chordotonal organ, J. Insect Physiology, 20:1219-1239.
Burns, M. D., and Usherwood, P. N. R., 1979, The control of walking
     in orthoptera: II Motor neuron activity in normal free-walking
     animal, J. exp. Biol., 79:69-98.
Burrows, M., 1973, Physiological and morphological properties of the
     metathoracic common inhibitory neuron of the locust, J. Comp.
     Physiol., 82:59-78.
Burrows, M., and Hoyle, G., 1973, Neural mechanisms underlying
     behaviour in the locust *Schistocerca gregaria*, II Topography
     of limp motoneurons in the metathoracic ganglion, J. Neurobiol.,
     4:167-186.
Campbell, J. I., 1961, The anatomy of the nervous system of the meso-
     thorax of locust migratoria migratoroides R and F, Proc. Zool.

Soc. Lond., 137:403-432.

Chopard, L., ed., 1951, "Faune de France," P. Lechevalier, Paris.

Evans, P. D., 1981, Multiple receptor types for octapamine in the locust, J. Physiol., 318:100-122.

Hoyle, G., 1974, A fuction for neurons (DUM) neurosecretory on skeletal muscle of insects, J. exp. Zool., 185:97-110.

Hoyle, G., 1978, Distribution of nerve and muscle fibre types in locust jumping muscle, J. exp. Biol., 73:205-233.

Hoyle, G., Dagan, D., Moberly, B., and Colqunhoun, W., 1974, Dorsal unpaired median insect neurons make neurosecretory endings on skeletal muscle, J. exp. Zool., 187:159-165.

Karnovsky, J. N., 1965, A formaldehyde-glutaraldehyde fixative of high osmolarity for use in electron microscopy, J. cell. Biol., 27:137

Theophilidis, G., and Burns, M. D., 1982, A gold-plated suction electrode for extracellular recording and dye infusion, J. exp. Biol., 98:455-457.

Usherwood, P. N. R., and Grundfest, H., 1965, Peripheral inhibition in skeletal muscle of insect, J. Neurophysiol., 28:497-518.

Usherwood, P. N. R., Runion, H. I., and Campbell, J. I., 1968, Structure and Physiology of a chordotonal organ in the locust leg, J. exp. Biol., 48:305-323.

Wilkens, C. A., and Wolf, G. E., 1974, A new electrode design for en passant recording, stimulation and intracellular dye infusion, Comp. Biochem. Physiol., 48A:217-220.

ADAPTATION INSECT - PLANT IN CYNIPID GALLS

R. Bronner

Laboratoire de Cécidologie (ERA 975 de l'Université
Strasbourg associée au C.N.R.S.)
28 rue Goethe, 67083 Strasbourg Cedex, France

INTRODUCTION

Plant galls constitute the most elaborately evolved plant -
insect relationships. In this paper some observations on galls
caused by Cynipids (Hymenoptera of the Cynipidae family) will be
presented, because the host relation of the Cynipids are intriguing
from the point of view of coevolution. Indeed, 80% of Cynipids
induce galls on Fagaceae, particularly on *Quercus* leaves, while
the remaining 20% are attached to only six plant families, mostly
to Rosaceae and Compositae (Dalla Torre and Kieffer, 1910; Kinsey,
1970). This distribution of Cynipid galls on a limited number of
plant families leads to a symbiotic adaptation between Cynipids
and their host-plants. The purpose of this paper is to define this
kind of adaptation.

IN A CYNIPID GALL, ALL THE GAINS ARE, AT FIRST SIGHT, IN BEHALF OF THE CECIDOZOA

Since the earliest developmental stage, a Cynipid gall, whatever its position on the host-plant is, induces sap flow disturbances and loss of substances for the plant. Indeed, it is well
known that galls are important sinks for substances coming from the
host-plant, and that sugars or phosphorus compounds supplied to
the plant are translocated and concentrated in galls (Fourcroy and
Braun, 1967; Jankiewicz et al., 1969).

More, according to its position on the plant, the gall provokes
several modifications more or less obvious, but always with a
negative aspect for the plant. So, a modification or an inhibition

61

of growth are observed in bud galls. Sometimes, the attacked bud
is entirely atrophied and replaced by a globular tissues mass.

Generally, the attacked plant produces great quantities of non-
essential tissues. Finally, the gall provokes a loss in seed yielding, when it develops in reproductive organs.

But, there is no observation showing that the Cynipid gall
development causes a modification which seems to be positive for
the host-plant. It is important to notice that Cynipid galls development, even in a large number, never causes death of the host-plant.

It is beyond doubt that the Cynipid, which induces a gall,
derives all the benefit. Indeed, beyond a protective device more
or less effective (Cosens, 1912; Fourcroy and Braun, 1967), the
larva receives all its nourishment upon a special gall tissue
called nutritive tissue.

Furthermore, sometimes, the gall assures the dispersal of the
species of the gall maker.

CYNIPID GALL SEEMS TO BE A PROTECTIVE DEVICE FOR THE PLANT IN
LIMITING THE CECIDOZOA INVASION

In most of Cynipid galls there is the same inner organization
scheme around the cecidozoa:
  - a larval cavity with the larva,
  - a nutritive tissue,
  - a lignified sheath,
  - a cortical tissue with numerous vascular bundles connecting to
    the vascular tissue of the plant,
  - an epidermis.

The nutritive tissue and the lignified sheath constitute the
"inner gall"; the "outer gall" is formed through the cortical tissue
and the epidermis.

Our observations on five Cynipid galls indicate that the size
of the "inner gall" is constant, whereas the size of the "outer
gall" varies for a given species (Fig. 1). This is illustrated on
Fig. 2 for *Liposthenes glechomae* L. gall on *Glechoma hederacea* L.,
where there are two galls of the same age, but with two different
sizes. The two different sizes are caused by a different development
of the "outer galls" when the "inner galls" have the same size.
Gaufilet and Fourcroy (1966) reported that in this gall the "inner
gall" has a spherical shape, with a constant diameter at 2.6 mm.
According to them, the lignified sheath is being found into a
parenchymatic tissue and always with an inner diameter constant
at 2.6 mm. It is well known that in Cynipid galls, there is a

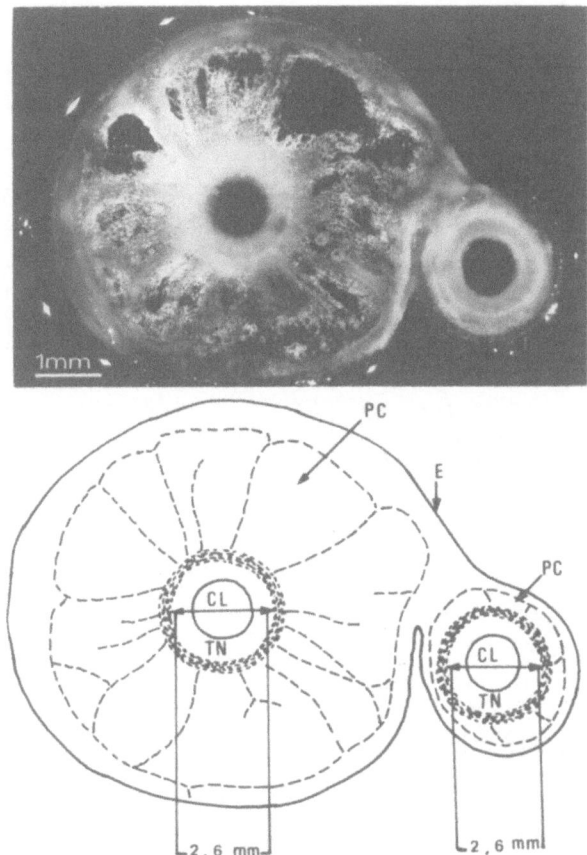

Fig. 1. Sizes of "inner galls" of four Cynipid galls.

progressive destruction of the nutritive cells by the larval feeding. At the end of its development, the larva is found near the
lignified sheath (Roth, 1949; Gaufilet and Fourcroy, 1966; Bronner,
1977). Thus, the mass of plant tissue used by the Cynipid larva for
its feeding is contant for a given species.

The problem is to know the decisive factor of this limit. Is
it the plant or the parasite? According to Cook (1923), the gall is
probably an "organ" developped by the plant to protect itself from
the injury of the cecidozoa, giving it a part of itself. It is
important to notice that this part, since the earliest stage of the
gall development is cut from the normal morphogenesis correlations
and is entirely controlled by the cecidozoa. Indeed, numerous
papers reported that the active presence of the larva is necessary
to bring up and to maintain the gall structure by acting on the

Biorhiza pallida Ol. ♀♂ on *Quercus.*

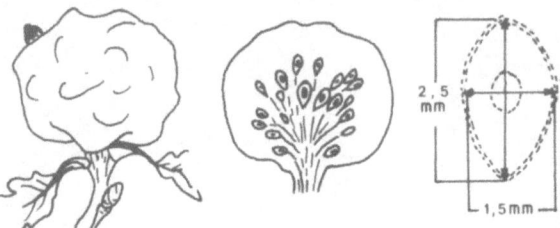

Neuroterus quercusbaccarum L. ♀♂
on *Quercus.*

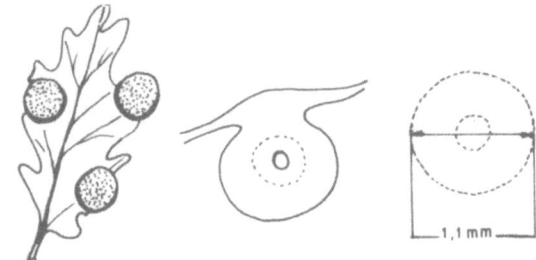

Andricus curvator Htg. ♀♂ on *Quercus.*

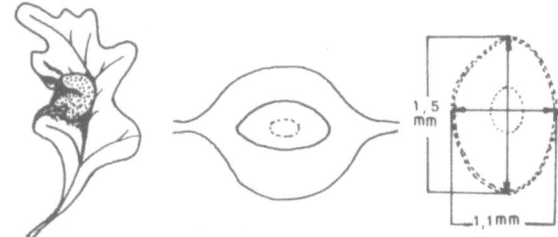

Diastrophus rubi Bché. on *Rubus.*

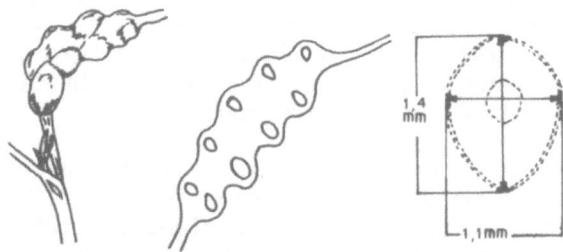

Fig. 2. *Liposthenes glechomae* L. gall on *Glechoma hederacea* L. leaves.
a) Section of two gall of the same stage and of two different
sizes: b) Interpretation of the section: the "inner galls"
are constant. (PC: cortical parenchyma; E: epidermis;
CL: larval cavity; TN: nutritive tissue).

cells of the gall. Any modification of the larval activity provokes a rapid regression of the gall structures even in the latest stages (Rohfritsch, 1971b, 1975; Bronner, 1977). In the case of Cynipid galls, Rohfritsch (1971a) experimentally showed that the formation and the development of the "inner gall" is strictly related with the larval activity, whereas the "outer gall" continues its development even after the removal of the larva. Thus, the "inner gall" is under the larval control, buth the "outer gall" depends on the plant.

THERE IS A COMPLETE HARMONY BETWEEN THE INSECT BIOLOGY AND THE GALL TISSUES FUNCTIONING

Cecidogenous Cynipid larvae feed on plant cells which are transformed into nutritive cells. Comparing with corresponding normal plant cells, and with other gall cells, the nutritive Cynipid cells exhibit typical cytological features (Molliard, 1897; Garrigues, 1951; Woll, 1954; Meyer, 1957; Rohfritsch, 1974; Rey, 1976; Bronner, 1977). These nutritive features are found in other insect galls, in Acarina, and Nematoda galls (Meyer and Maresquelle, 1983). In fact, they are found in different normal plant cells which are known to have high physiological and metabolical activities (companion cells of phloem: Essau, 1972; storage cells of cotyledons in seedlings: Öpik, 1964; secretory cells of nectary glands: Zandonella, 1970).

Nutritive Cynipid cells are rich in soluble compounds like sugars and amino-products (Bronner, 1977). These soluble compounds can be a product of the host-plant or are produced through the activity of hydrolases, which are very active in the nutritive cells (Bronner, 1977). Furthermore, nutritive cells are rich in starch and lipids. contents (Bronner, 1977). Starch is found in the cell-layers away from the larva and is transformed into soluble sugars during the gall development. Lipids are concentrated in the cells nearest to the larva and they probably serve also as food, because Roth (1949) succeeded in nourishing Cynipid larvae with milk. During their feeding, the Cynipid larvae progressively destroy all the nutritive cells. Bronner (1977) reported that this destruction is a result of a double process: first, a lysis of nutritive cells whose the walls remain intact. This lysis, result of a lysosomal process, is probably caused by proteasic enzymes of the larva. In a second phase, the larva acts with its strong mouth-parts (Roth, 1949).

In the larval cavity, the Cynipid larva is weltering in a liquid coming from the destroyed cells. But this larva does not excrement in its nutritive liquid, because, as in the most Hymenoptera, the mid-gut and the post-gut are without connection in the larval stage (Roth, 1949; Grassé et al., 1961). Unlike in other insect galls. Cynipid galls are entirely closed, the larva occupies

all the volume of the larval cavity and the increase in its size
follows the increase of the larval cavity.

CONCLUSIONS

Cynipid galls show a good example of a perfect coordination
between the gall-inducing organism and the gall-bearing plant.
Zweigelt (1942) has developed the idea that the formation of a
gall is the result of a gradual process of mutual adaptation between the parasite and the plant. According to Mani (1964), "the
primary reaction of the plant to the gall inducing organism is not
in the interest of the organism, but represents a sort of struggle
against the attack of the parasite". But, if it is true, the
parasite is now well adapted to live in the gall.

In the meantime, this adaptation forced the parasite to extreme
specialization. One example of this specialization can be found in
the capacity to lyse plant cells of Cynipid larvae and also of
Cynipid eggs. Indeed, the first stage of Cynipid galls is a lysis
of the plant tissues in contact with eggs which is caused by enzymes
(e.g. proteases) produced by these same eggs (Bronner, 1973). The
strong specificity between Cynipid and host may be explained by
this lytic capacity. Thus, it is known that plants have protease
inhibitors, which act against insect attacks (Green and Ryan, 1972;
Ryan, 1973). It is possible that in host-plant of cecidogenous
Cynipids, these protease inhibitors may be:
  - normally absent,
  - inhibited by the parasite, or
  - destroyed by the parasite.

REFERENCES

Bronner, R., 1973, Propriétés lytiques des oeufs de *Biorhiza pallida*
    L. ♀♂, C.R. Acad. Sci., Paris, 276, série D:189-192.
Bronner, R., 1977, Contribution à l'étude histochimique des tissus
    nourriciers des zoocécidies, Thèse, Strasbourg(1976), Marcellia,
    40:1-34.
Cook, M. I., 1923, The origine and structure of plant galls, Science,
    26:6
Cosens, A., 1912, Acontribution to the morphology and biology of
    Insect galls, Transactions of the Canadian Institute, vol. IX:
    297-387.
Dalla Torre, K. W. von and Kieffer, J. J., eds., 1910, "Cynipidae,
    Das Tierreich," 24ste Lieferung, Friedlän. u. Son Verlag, Berlin.
Essau, K., 1972, Cytology of sieve elements in minor veins of sugar
    beet leaves, The New Physiologist, 71:161-168.
Fourcroy, M., and Braun, Ch., 1967, Observations sur la gall de l'

Aulax glechomae L. sur *Glechoma hederacea* L. II. Histologie et rôle physiologique de la coque lignifiée, Marcellia, 34:3–30.

Garrigues, R., 1951, Recherches sur les cécidies, le cancer et l' action des carbures cancérigènes sur les végétaux, Thèse, Lyon.

Gaufilet, C., and Fourcroy, M., 1966, Observations sur la galle de l'*Aulax glechomae* L., I, Structures et croissance, Marcellia, 33:3–31.

Grassé, P. P., Poisson, R., and Tuzet, O., eds., 1961, "Zoologie, I, Invertébrés, Précis de Sciences Biologiques,"Masson et Cie., Paris.

Green, T. R., and Ryan, C. A., 1972, Wound-induced proteinase inhibitor in plant leaves. A possible defence mechanism against insects, Science, 175:776–777.

Jankiewicz, L. S., Plich, H., and Antoszewski, R., 1969, Preliminary studies on the translocation of ^{14}C-labelled assimilates and [32]PO$_3$ towards the gall evoked by *Cynips* (*Diplolepis*) *quercusfolii* L. on oak leaves, Marcellia, 36:163–174.

Kinsey, A. C., 1920, Phylogeny of Cynipid genera and biological characteristics, Bull. Amer. Mus. nat. Hist., 42:357–402.

Mani, M. S., ed., 1964, "Ecology of plant galls, Monographiae Biologicae,"Volumen XII, Dr. W. Junk Publishers, The Hague.

Meyer, J., 1957, Cécidogenèse comparée de quelques galles d'Arthropodes et évolution cytologique des tissues nourriciers, Thèse, Strasbourg.

Meyer, J., and Maresquelle, H. J., eds., 1983, "Anatomie des galles, Handbuch der Pflanzenanatomie," Bornträger Ed,. Stuttgart.

Molliard, M., 1897, Hypertrophie pathologique des cellules végétales, Rev. Gén. Bot., 9:33–45.

Öpik, H., 1964, The form of nuclei in the storage cells of the cotyledons of germinating seed of *Phaseolus vulgaris* L., Exp. Cell Res., 38:517–522.

Rey, L., 1976, Recherches comparées sur trois types d'Hyménoptérocécidies (Cynipides et Tenthrédinides). Etude morphologique, hostologique et ultrastructurale du développement et de la différenciation. Essais expérimentaux, Thèse, Nantes.

Rohfritsch, O., 1971a, Développement cécidien et rôle du parasite dans quelques galles d'Arthropodes, Thèse, Strasbourg (1970), Marcellia, 37:233–339.

Rohfritsch, O., 1971b, Culture in vitro de jeunes galles d'*Aulax glechomae* L. sur *Glechoma hederacea* L., Marcellia, 37:151–161.

Rohfritsch, O., 1974, Infrastructure du tissu nourricier de la galle de l' *Aulax glechomae* L. sur le *Glechoma hederacea* L., Protoplasma, 81:205–230.

Rohfritsch, O., 1975, Etude comparative de cellules du tissu nourricier de la jeune galle de l'*Aulax glechomae* L. sur le *Glechoma hederacea* L. et des cellules du tissu nourricier abandonné par le parasite, Marcellia, 38:185–196.

Roth, P., 1949, Beiträge zur Biologie der Gallwespen, Verh. Naturf. Ges. Basel, 60:104–178.

Ryan, C. A., 1973, Proteolytic enzymes and their inhibitors in plants,

        Ann. Rev. Plant Physiology, 24:173-196.
Woll, E., 1954, Beiträge zum Differenzierungsproblem an Hand der
        Zytologie von Pflanzengallen, Z. Bot., 42:1-29.
Zandonella, P., 1970, Infrastructure des cellules du tissue nectari-
        gène floral de quelques Caryophyllaceae, C. R. Acad. Sci.,
        Paris, 270, Série D:1310-1313.
Zweigelt, F., 1942, Beiträge zur Kenntnis der Blattlausgallen, Biol.
        Gen. 16(4):554-572.

ADAPTATION OF GALL MITES (ACARI, ERIOPHYOIDEA) TO LIVE IN GALLS

E. Westphal

Laboratoire de Cécidologie, Equipe de Recherche de
l'Université Strasbourg I associée au C.N.R.S. (ERA 975)
Institute de Botanique
28 rue Goethe, 67083 Strasbourg Cedex, France

In Europe, more than 350 mites species induce galls on plants
(Buhr, 1964) and these species belong to the family Eriophyoidea.
These gall mites are almost invisible (0,1-0,3 mm long to the un-
aided eye) so that the identification of their galls were at first
only based on the host plant (Frauenfeld, 1864; Löw, 1874). Nalepa
(1911) contributed to Eriophyoidea taxonomy which was recently reviewed by Keifer (in Jeppson et al., 1975).

Although entirely phytophagous, the Eriophyoid family includes
both gall mites and open leaf vagrants or rust mites which cause
more or less important damage to their hosts. The soft wormlike
body of Eriophyoids has the three standard acarine sections (Baker
and Wharton, 1952): (1) gnathosoma or rostrum; (2) propodosoma
showing dorsally the cephalothoracic shield and ventrally the
insertion of the two pairs of legs; (3) hysterosoma with transverse
rings (tergites and sternites).

Compared to other phytophagous Acari (Tetranichidae, Tarsen-
omidae, etc., all bearing 4 pairs of legs) the great reduction in
Eriophyoid body structure, the loss of the rear 4 legs and nearly
all body setae have been interpreted (Jeppson et al., 1975) as an
adaptation to live in restricted microenvironments on plants (bark
crevices, bases of outer bud scales or galls). However, particular
adaptations to live in galls are manifested by: (1) the reduction
of the anterior shield extension, (2) a more wormlike body with
narrow rings heavily microtuberculate (Jeppson et al., 1975). In
contrast, free-living species entirely exposed to fluctuations of
external conditions have larger and less tuberculated tergites.
Existence of *Paraphytoptus* spp. constitutes an intermediated form

69

between open leaf mites and species living under cover: species of this genus do not induce galls but live partially protected in natural plant hair masses. Only the anterior part of the hysterosome protected by these hairs offers the narrow heavily microtuberculate rings. In contrast, the exposed rear part has large tergites with reduced microtuberculation. Thus, only life in extremely protected niches (hair masses or galls) removing mites from direct contact with sunlight and atmospheric fluctuations permits Eriophyoids to conserve an important abdominal ring micro-tuberculation. Such an adaptive interpretation may be confirmed by the morphological variations (deuterogyny or alternation to genera-tions) in gall mites living on deciduous host-plants (Jeppson et al., 1975). Galls are initiated by protogyne females and when galls be-come mature, they are gradually filled with deutogyne females which are more deeply pigmented and less microtuberculate. Summer and winter females of the same species may differ sufficiently enough to be considered as different genera. Deutogynes leave the galls, hibernate and in spring they travel to select new organs for gall building. The great reduction of the abdominal microtuberculation shown by deutogynes constitute an adaptation to survive unfavourable periods outside galls since it was established that ring microtuber-cles have a function in water regulation through mite skin (Jeppson, et al., 1975).

The present study reports on the structure of gall mite mouth-parts and on feeding punctures left on the host plant. Two examples have been chosen: (1) *Eriophyes cladophthirus* Nal. inducing witches' brooms and virescences on *Solanum dulcamara* L., and (2) *Phytoptus padi* Nal. inducing pouch galls on the leaves of *Prunus padus* L.

1) Mouth parts

The strongly downcurved rostrum (Fig. 1) is enclosed on each side by two pedipalps composed of four telescoping segments. Its terminal part forms a suction cup (6 μm in diameter) which adheres to the host cell during penetration of the chelicerae. The rostrum is about 20 μm long and holds five mouth stylets in an anterior groove (Figs. 2 and 3). The two anterior stylets (or chelicerae) are separate for their entire length (18-20 μm) but lie close together, without making any salivary or alimentary duct. They move alternately and are forced on a short distance into the host cell wall. Behind the chelicerae there are two auxiliary stylets and one oral stylet. The precise function of these three stylets is not clear and needs further investigation. Our observations confirm those previously related on different gall mite species by Keifer (1959) and Krantz (1973), using ordinary light microscope. The fundamental structure of Eriophyoid mouthparts is similar for gall mites and free living forms, and seems well adapted to phytophagous life.

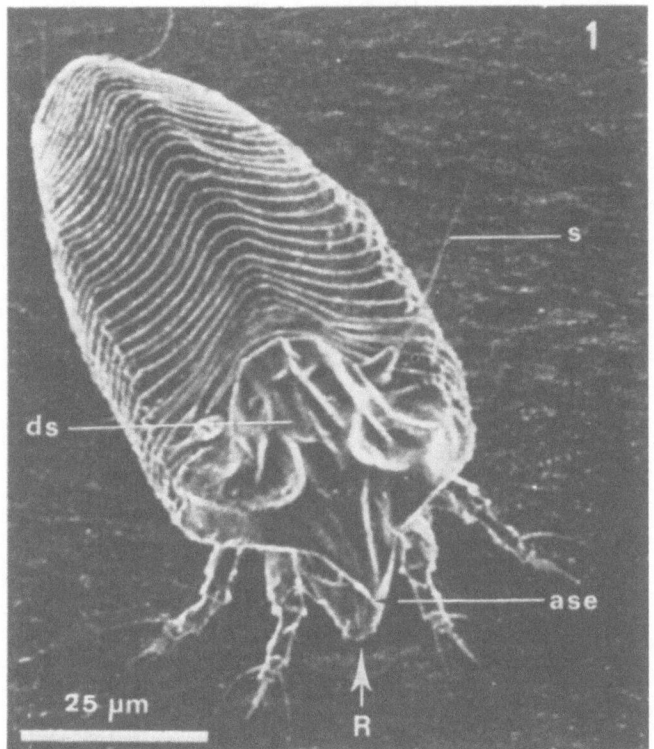

Fig. 1. Feeding stance of *Eriophyes cladophthirus*. Scanning Electron
        Microscope view.
        ds: dorsal shield; ase: anterior shield extension; R: rostrum;
        s: anterior dorsal shield seta.

2) <u>Feeding punctures</u>

        Light microscopic observations (Němec, 1924) and ultrastruc-
tural studies (Westphal, 1977) have shown that gall mites always
leave feeding punctures in the host tissues. The cell wall per-
foration during mite feeding provokes the gradual formation of a
cone-shaped wall thickening (Fig. 4) which appears as a typical
disk-like structure in cross section (Fig. 5). Callose deposits
confer an heterogenous aspect on feeding punctures (Westphal, 1977).
The persistence of these structures is one of the most original
feature of mite galls. However, Eriophyoid living forms as *Abaccarus
hystrix* (Gibson, 1974) also leave feeding punctures on their host
plants. Scanning electron microscope observations permit to suppose
that these feeding punctures are similar.

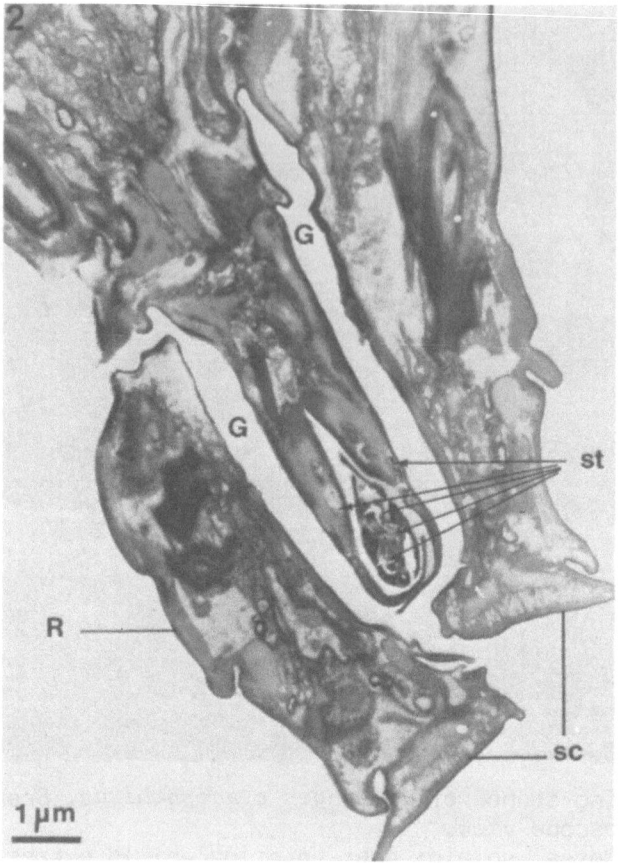

Fig. 2. Longitudinal section of the mouthparts of *E. cladophthirus*.
G: grove; R: rostrum; sc: suction cup; st: stylets.

For mite inducing galls, important cytological changes occur simultaneously with the feeding puncture elaboration. These modifications lead to the differentiation of nutritive cells (Westphal, 1977) similar to those described in insect or nematode galls (Meyer, 1982). The nucleus and nucleolus hypertrophy, the chromatin dispersion, the cytoplasm enrichment, the high density of organelles (ribosomes, mitochondria, etc.), the vacuole fragmentation, occur within the first hour of gall mite attack (Westphal, 1982) and express features of high metabolic activity. Some hours later, cellular divisions and growth of adjacent tissues lead to gradual gall formation where gall mites will find the sole microenvironment suitable for their reproduction. In contrast after the puncture of

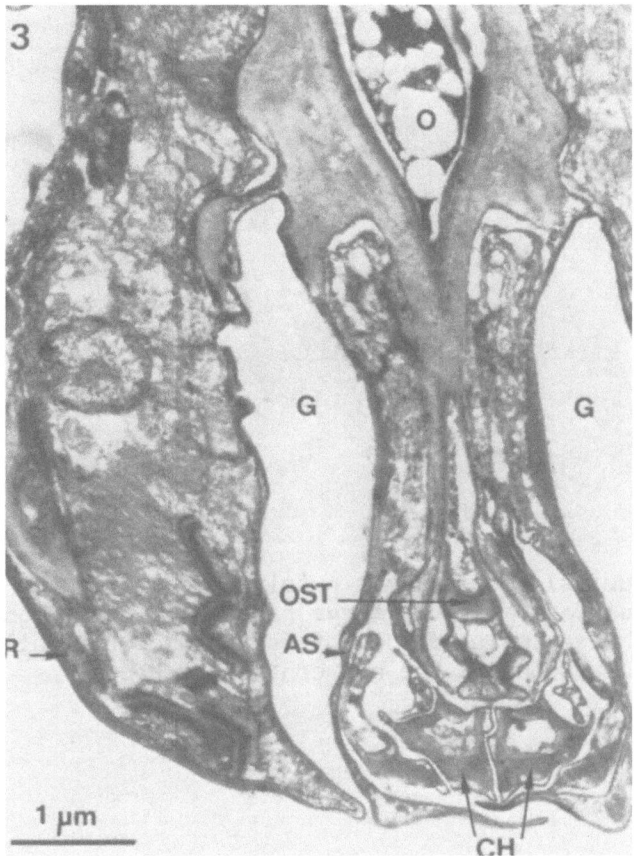

Fig. 3. More oblique section of the mouthparts.
      AS: auxiliary stylets; CH: chelicerae; G: grove; O: oeso-
      phagus; OST: oral stylet.

free living species, the injured cells do never show such cytological changes (Westphal, 1977).

The fact that only gall mites are able to induce typical nutritive cells constitute a good evidence of adaptation not only of
gall mites themselves, but also of host plants that specifically
react by gradual cecidogenesis leading to galls of perfectly well
defined shape and color.

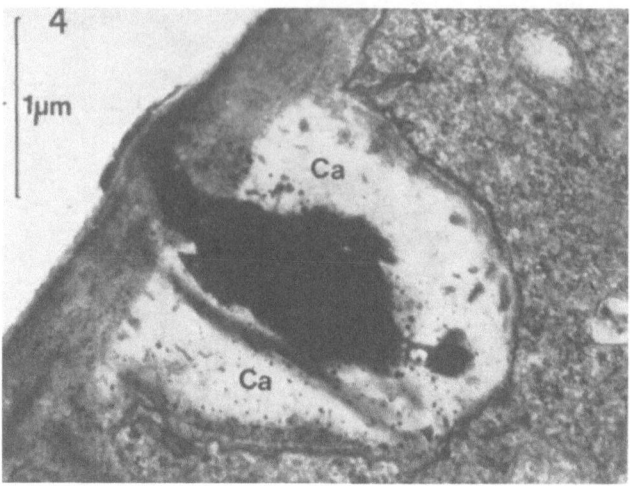

Fig. 4. Longitudinal section of a nutritive cell (in the gall of
Phytopus padi Nal. on *Prunus padus* L.) with a cone-shaped
wall thickening around the feeding puncture. Callose deposits
(Ca) correspond to the non contrasted areas.

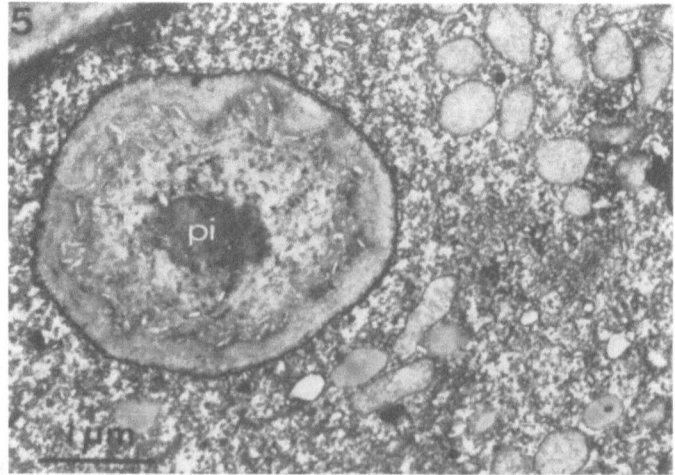

Fig. 5. Cross section of a feeding puncture showing a disk-like
structure (arrow). Note the abundance of organelles. Gall
of a *P. padi* Nal.

REFERENCES

Baker, E. W., and Wharton, G. W., eds., 1952,"An introduction to
    Acarology", The Mac Millan Co., New York.
Buhr, H., ed., 1964/1965, Bestimmungstabellen der Gallen an Pflanzen
    Mittel- u. Nordeuropas, I and II, Fischer, Jena.
Frauenfeld, G., 1864, Beitrag zur Metamorphosen. Geschichte aus dem
    Jahre 1862,III, Verschiedene Metamorphosen, Verhandl. d.k.k.
    Zool. Bot. Gesellschaft (Wien) 13:1230-1236.
Gibson, R. W., 1974, Studies on the feeding behaviour of the erio-
    phyid mite Abaccarus hystrix, a vector of grass viruses, Amn.
    appl. Biol., 78:213-217.
Jeppson, L. R., Keifer, H. H., and Baker, E. W., eds., 1975,"Mite
    injurious to economic plants", University of California Press,
    London.
Keifer, H. H., 1959, Mouthparts of eriophyid. Eriophyid studies,
    XXVI, Bull. Calif. Dept. Agr., 47:278-281.
Krantz, G. W., 1973, Observations on the morphology and behaviour
    of the filbert rust mite Aculus comatus (Prostigmata: Erio-
    phyoidea) in Oregon, Ann. Ent. Soc. Am., 66:709-717.
Löw, F., 1874, Über Milbengallen (Acarocecidien) der Wiener Gegend,
    Verhandl. Zool. Bot. Gesellschaft (Wien) 24:495-508.
Meyer, J., ed., 1982,"Anatomie des galles", Bornträger, Stuttgart
    (in press).
Nalepa, A., 1911, Eriophyiden, Gallmilben, Zoologica XXIV, 61:166-
    293.
Němec, B., Untersuchung über Eriophyidengallen, Studies from the
    Plant Physiological Lab. of Charles Univ., Prague, V, 2:47-94.
Westphal, E., 1977, Morphogenèse, ultrastructure et étiologie de
    quelques galles d'Eriophyides (Acariens), Marcellia, 39:193-375.
Westphal, E., 1982, Cellular dedifferentiation in leaf epidermis
    during early stage of mite gall cecidogenesis, XVth EDBO Inter-
    national Embryological Conference.

DIVERSITY AND UNITY BY ARTHROPOD GALLS. AN EXAMPLE: THE BUD GALLS

F. Dreger-Jauffret

Laboratoire de Cécidologie (ERA 975 de l'Université
Strasbourg I associée au C.N.R.S.),
28 rue Goethe, 67083 Strasbourg Cedex, France

INTRODUCTION

A gall is a deformation of a plant provoked by a cecidogenetic
agent. This deformation is strictly correlated to the presence of
the gall maker. Sometimes we can observe only inhibition phenomena
(curling, wrinkling of leaves, or axis). Most often complex phenomena of hyperplasia and hypertrophia involve the edification of
specific morphogenesis resulting in galls.

The perfect adaptation between the gall maker and its host
plant leads to a very large variation in forms and mechanisms.

It is in Arthropods and especially Acarina and Insects that we
can find most of gall makers. Amongst Insects, Thysanoptera, Homoptera, Diptera and Hymenoptera are mainly responsible for the plant
galls. We can also find some interesting gall makers in Coleoptera
and Lepidoptera.

Among host plants, differences of liability are visible according to families. Houard (1908/1909/1913, 1922/1923, 1933, 1940),
author of a lot of catalogues describing the galls of Europe, Asia,
Africa, North and South America, counted 500 galls on 150 plants
of the family Salicaceae. According to Mani (1964) 50% of the galls
in Europe and North America are induced on Fagaceae.

Distribution of galls also varies according to the geographical conditions. Thus, in Euphorbiaceae, we can find 120 galls in Asia,
49 galls in Europe and only 6 galls in North America (Felt, 1940).
In Australia, Myrtaceae are predominant and 50% of known galls in
Europe and 50% of known galls are reported in this family (Mani,
1964).

At last, special affinity between some families should be emphasized: 80% of the Cynipid galls occur on Fagaceae, and all Tenthredinid (Hymenoptera) galls are found on Salicaceae.

Galls are found on all plant organs. According to families of gall makers, we can note some affinity with such or such part of plants. For instance, the Thrips galls especially occur on leaves, and the Cecidomyid galls are very rare on roots, but Coleoptera galls are found, as a rule on stem and root.

GALLS ARISING ON THE TIP OF THE SHOOT, THE BUD AND THE APEX

Galls arising on the vegetative point of the plant show various aspects and development, but they all have two types of effects in common.
a) The so called "organoid effects" provoke deformations of plant organs, which remain recognizable.
b) The so called "histoid effects" provoke a total transformation of organs or induce new specifically cecidian structures. For instance, the nutritious layers or the lignified sheath are typically histoid effects.

In the case of bud galls, organoïd effects are especially important:
- stopping of the growth of the axis;
- shortening of internodes;
- enlargement of the apex;
- effects on leaves (multiplication, deformation...).

Histoid effects moreover concern the transformation of meristematic tissues into nutritious tissues (Dreger-Jauffret, 1978).

According to Mani (1964) Diptera and Hymenoptera are responsible for the 40% of bud galls, respectively, while Lepidoptera and Coleoptera cause less than 10% of bud galls, each.

In this chapter we will see different types of galls where the parasite remains external to the apex or becomes internal. Special emphasis will be paid on witch brooms, provoked by Arthropods.

Galls provoked by enternal parasite, affecting apex or bud

"Rosette" or "artichoke" galls
Taxomyia taxi Inch (Cecidomyid) on *Taxus baccata* L. (Fig. 1).

Many authors have worked on this gall (Houard, 1904; Lovett and Chappell, 1970; Jauffret et al., 1970; Dreger-Jauffret, 1978). Eggs are laid at the end of May. The new hatched larvae crawl along the young twig and creep into the young bud. But only one larva remains on the vegetative point. During the first stage of the two

years cycle, this larva induces an enlargement of the apex. Nevertheless, the apex continues its function and gives many leaves without sensible lengthening of the axis. This function results in a
small artichoke. During the 2nd year, the larva grows up, the apical
meristem is thoroughly modified, transformed into nutritive pad and
it lost its organogenetic properties. When the larva pupates, the
nutritious tissue decays and becomes parenchymatous. At the end of
May, the image escapes, leaving the empty pupa at the top of the
artichoke, which can persist intact for some years.

   Bayeria capitigena Bre. (Cecidomyid) on *Euphorbia cyparissias*
L. (Fig. 3).

   This gall develops quickly, within some weeks on an herbaceous
weed (Houard, 1904; Jauffret, 1973; Dreger-Jauffret, 1978); the
larvae are numerous on the same apex and they soon provoke severe
transformations of the vegetative point. After the production of enlarged leaves (often brightly colored), which constiture the rosette,
the morphogenetic activity of the apex rapidly decreases. The last
formed leaves are very modified, without chlorophyll and often
partially transformed into nutritious leaves. The apex is also transformed into nutritious tissue and some necroses can be seen. Larvae
pupate inside the galls; when they escape the gall decays and dies.

Fig. 1. Galls of *Taxomyia taxi* Inch on *Taxus baccata* L.
        (A) Two bud galls at the stage when the "artichoke" is form-
        ing (1st year of cecidogenesis); (B) Longitudinal axial
        section accross a mature gall, showing the larval room and
        the apical meristem transformed into nutritive pad.

Fig. 2. Mature gall of *Bayeria capitigena* Bre. on *Euphorbia cyparis-
        sias* L. (the arrow shows the left pupae after the hatching
        of the midges).

These tow galls moreover differ by the number of larvae and
the duration of their life cycle.

"Big buds"
Cecidophyopsis psilaspis Nal. on *Taxus baccata* L. (Fig. 3).

We chose the example of *Cecidophyopsis psilaspis* Nal. since
it developes on *Taxus baccata* L., allowing thus the comparison
between the effects caused by a Cecidomyid and an Acarina on the
same host plant (Jauffret and Westphal, 1974; Westphal, 1976; Dreger-
Jauffret, 1978). According to the number of parasites, the bud is
more or less attacked, and in most cases the whole bud is deformed.

The Acarina provoke an inhibition of the internodal growth and
stop the organogenetic function of the apical meristem. At the
beginning of the attack, the apex is quickly transformed, some parts
proliferate in nutritious cells, while some others show necroses.
The same is happening with the very young leaves. In this way Acarina give us an example of a more destructive cecidogenesis than
Cecidomyids.

Galls with internal parasite

As a rule, buds with internal parasites are more transformed
than those with only external parasites. The internal position of
the larva can result of the very oviposition, or of ulterior growth
of tissues surrounding the larva.

 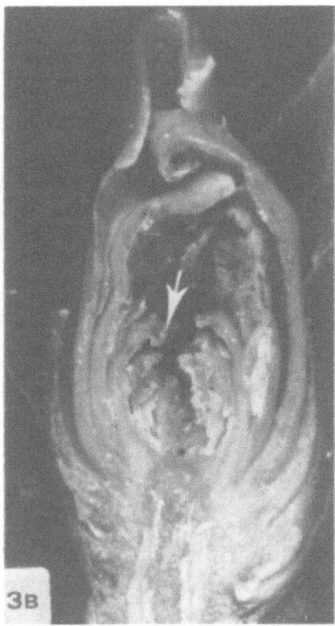

Fig. 3. Galls of *Cecidophyopsis psilaspis* Nal on *Taxus baccata* L.
(A) The arrow shows the bud parasitized by mites. Other
buds are healthy; (B) Longitudinal axial section accross
a gall showing the young leaves deformed and covered with
nutritive cells forming outgrowth (arrow).

Andricus fecundatrix Hartg. ♀♀ (Hymenoptera, Cynipidae) on
Quercus pedunculata Ehrh. (Fig. 4).

As other Cynipidae, the biological cycle of *A. fecundatrix*
involves alternating generations throughout the year. The sexual
generation gives a spring gall on male flowers. In June, only
females spring out of this gall, and lay their eggs into the young
buds. These ones build artichoke-like galls with scaly leaves.
After this stage the larvae embed themselves through lysis into
the tissues of the bud, the function of which is then interrupted.

Exhibiting the specific capacities of Cynipidae, the larva
builds up the so called "internal gall" (involving nutritive tissue,
reserve tissue and sclerified sheath) which can fall to the ground,
being by this way protected during the winter.

Lipara lucens Magnus (Diptera, Chloropidae) on *Arundo phragmites* L. (Figs. 5 and 6).

The eggs are laid on leaves and the new hatched larvae make

 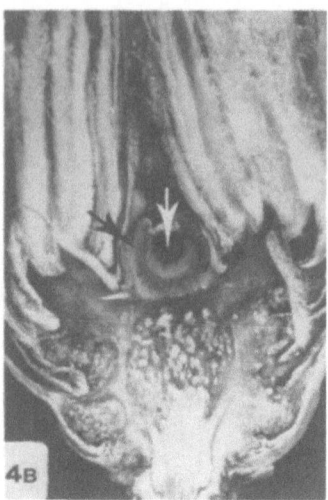

Fig. 4. Galls of *Andricus fecundatrix* Hartg. ♀♀ on *Quercus pendun-
culata* Ehrh. (A) Three adult artichoke-like galls (on the
lower face of the leaves three kinds of *Neuroterus* "lentil
galls" can be seen); (B) Longitudinal axial section accross
a mature gall.
- the white arrow shows the larval room surrounded by the
nutritive cells, light colored.
- the black arrow shows the lignified sheath.

their way to the bud. The apex then enlarges, gives larger leaves
and the internodal elongation stops. Then the larvae consume the
meristematic tissues and dig up a hole along the stem (Ross, 1932;
Ruppolt, 1959) into which they will pupate.

   Harmolitha hyalipenne hyalipenne Walck (Hymenoptera, Chalcidi-
dae) on *Agropyrum repens* P.B. (Fig. 7).

   According to Docters van Leeuwen (1907), the eggs are laid
into the stem, near the apex. The larva induces the formation of
a nutritive tissue with an external sclerified sheath. The apex is
enlarged and gives leaves with reduced lamina and enlarged sheath.
The internodes are reduced too. When the gall is mature, the apex
decays and an axillary bud often appears.

Galls forming witch-brooms

   This type of galls is due to the anarchic proliferation of
axillary buds and presents then a preponderance of organoid effects.

   Clinodiplosis graminicola Kief. DVL. (Diptera, Cecidomyidae)

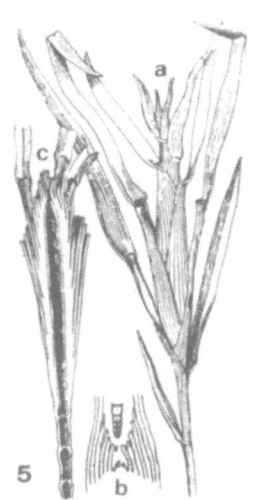 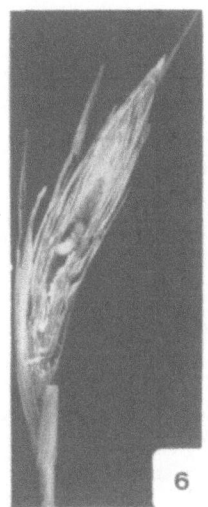 

Fig. 5. Gall of *Lipara lucens* Magnus, on *Arundo phragmites* L.
(a) General view of a mature gall; (b) Schematic longitudinal axial section accross a young gall; (c) Schematic longitudinal axial section accross an adult gall (after Ross,
1932).

Fig. 6. Longitudinal axial section accross a mature gall.

Fig. 7. Gall of *Harmolita hyalipenne hyalipenne* Walk on *Agropyrum
repens* P.B. (a) General view of an adult gall; (b) Schematic longitudinal section accross an adult gall (after
Houard, 1904).

on *Cynodon dactylon* Pers. (Fig. 8).

According to Docters van Leewen (1926) the eggs are laid on
the leaves. New hatched larvae creep into the young leaves. Under
their action the apex gives rise to many axillary buds.

During a second stage, each larva settles itself in an axillary
bud; these axillary buds then develop into shoots which form the
typical witch-broom.

Livia juncorum Latr. (Homoptera, Psyllidae) on *Juncus articulatus* L. (Figs. 9 and 10).

The gall is due to the activity of many larvae and its degree
of development varies with the number of parasites. According to
Schmidt and Meyer (1966) eggs with a short peduncle are inserted on
the edge of the leaves. The new hatched larvae creep between the
leaves of vegetative or floral buds. The stem then becomes thicker.

8

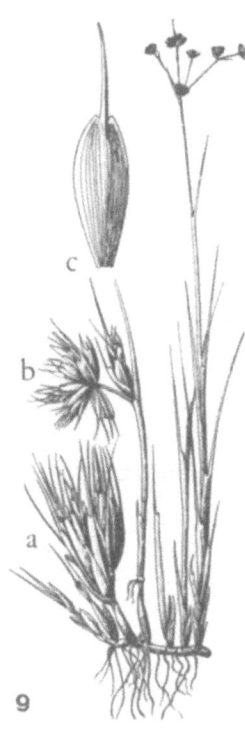

9

10

Fig. 8. Adult gall forming a witch broom provoked by *Clinodiplosis graminicola* Kief. & DVL on *Cynodon dactylon* Pers (after DVL & Reijnvaan, 1926).

Fig. 9. Galls of *Livia juncorum* Latr. on *Juncus articulatus* L. (a) Gall on vegetative bud; (b) Gall on floral bud; (c) deformed leaf showing the enlarged sheath and the reduced lamina (after Ross, 1932).

Fig. 10. Adult galls on floral buds (photo by Westphal).

The internodes shorten and new axillary buds appear, which give short leaved shoots.

Gall due to an unknown Acarina on *Carmichaelia violacea* (Fig. 11).

Mites proliferation provokes on this thread-like plant of the pulse family spectacular cluster of little leaved shoots like those growing in the wet season.

Here the cecidogenetic effect is strictly limited to the correlations between buds and shoots, which are locally perturbed, to give a gall offering a shelter to the mites in this specific (very dry) climate.

Aceria geranii Can. (Acarina) on *Geranium* sp. (Fig. 12).

Houard (1904) had shown the result of the action of mites on this plant concerning its habits (internodes shortened and thickened) and the aspects of leaves (reduced and wrinkled). Diseased parts of the plant are covered with short hairs. Besides the local effects, Houard emphasizes the remote effect of the parasite on the apex of the plant.

Fig. 11. Witch broom due to an unknown Acarina on *Carmichaelis violacea* Kirk. (after Arnold, 1967).

Fig. 12. Witch brooms due to *Aceria geranii* Can. on *Geranium sanguineym* L. Note the shortened internodes and the proliferation of deformed leaves (after Schlechtendal, 1916).

CONCLUSION

These few examples allow the comparison between the action of different kinds of cecidozoa on similar target: the vegetative point. We can observe that it is the biology of cecidozoa, their gentle or savage feeding, movements and behavior of their living as well as their number too, that give rise to the different kinds of adaptations. *Taxomyia taxi* gives an example of a Cecidomyid with a long cycle, living on a tree, an unique larva makes a gall, little destructive, remaining several years in place. In a contrast, *B. capitigena*, another Cecidomyid, has a short biological cycle, numerous larvae live on an herbaceous plant and provoke both organoid and destructive effects (building of the rosette, and presence of necrotic places besides the nutritive cells).

In the case of witch brooms, organoid effects are predominant, a large variety of insects give rise to these galls, and this is a proof of the great adaptability of cecidozoa to the various environmental conditions.

REFERENCES

Arnold, B. C., 1967, A hitherto unrecorded mitegall on *Carmichaelia violacea* Kirk., Adv. Front. of Plant Sciences, 18:1-6.
Docters van Leeuwen-Reijnvaan, W. & J., 1907, Über die Anatomie und die Entwicklung einiger *Isosoma* Gallen auf *Triticum repens* und *Junceum* und über die Biologie der Gallenformer, Marcellia, 6: 68-101.
Docters van Leeuwen-Reijnvaan, W. & J., eds., 1926, "The zoocecidia of the Netherlands East Indies," Drukkerij De Unie, Batavia.
Dreger-Jauffret, F., 1978, Effets du parasitisme sur le méristème apic apical et le développement de quelques plantes ligneuses et herbacées, Thèse, Strasbourg.
Felt, E. P., ed., 1940, "Plant galls and gall makers," Comstock Publish. Cie.
Houard, C., 1904, Recherches anatomiques sur les galles de tiges: Acrocécidies, Ann. Sc. Nat. 20(19):289-38
Houard, C., ed.,   /1909/1913, "Les zoocécidies des plantes d' Europe et du Bassin de la Méditerranée," 3 vol., Herrmann et Cie., Paris.
Houard, C., ed., 1922/1923, "Les zoocécidies des plantes d'Afrique d'Asie et d'Océanie," 2 vol., Hermann et Cie., Paris.
Houard, C., ed., 1933, "Les zoocécidies des plantes de l'Amérique du Sud et de l'Amérique Centrale," Hermann et Cie., Paris.
Houard, C, ed., 1940, "Les zoocécidies des plantes de l'Amérique du Nord -Galles de Chênes-," Hermann et Cie., Paris.
Jauffret, F., 1973, Etude ultrastructurale de la transformation du méristème apical de l'*Euphorbia cyparissias* L. en tissu nourricier sous l'action du *Bayeria capitigena* Bre., C.R. Acad.

Sci., Paris, 276:1177-1180.

Jauffret, F., Rohfritsch, O., and Wasser, Ch., 1970, Observation sur l'action cécidogène de *Taxomyia taxi* Inch. sur le bourgeon de *Taxus baccata* L. et la transformation du méristème apical en tissu nourricier, C. R. Acad. Sci., Paris, 272:1767-1770.

Jauffret, F., and Westphal, E., 1974, Etude ultrastructurale de la transformation du méristème apical du *Taxus baccata* L. sous l'action d'un Acarien, le *Cecidophyes psilaspis* Nal., C. R. Acad. Sci., Paris, 278:1769-1772

Lovett, T. J., and Chappell, H. G., 1970, A note on the life cycle of the gall midge *Taxomyia taxi* Inch. on *Taxus baccata* L., Marcellia, 36(4):257-260.

Mani, M. S., ed., 1964, "Ecology of plant galls, Monographiae Biologicae," volumen XII, Dr. W. Junk Publishers, The Hague.

Ross, H., 1932,"Praktikum der Gallenkunde (Cecidologie)," W. Schoenichen, ed., Julius Springer Verlag, Berlin

Ruppolt, W., 1959, "Harzgallen" an Kiefern, Orion Z. Nat. u. Techn., 8:596-599.

Schlechtendal, D. H. R., 1916, Eriophyiden-cecidien, die durch Gallmilben verursachten Pflanzengallen, Zoologica, 24(61), Stuttgart.

Schmidt, E., and Meyer, J., 1966, Observation sur la structure des galles de *Livia juncorum* sur *Juncus articulatus* L., Marcellia, 33(4):237-253.

Westphal, E., 1976, Morphogenèse, ultrastructure et étiologie de quelques galles d'Eriophyidae (Acariens), Thèse, Strasbourg.

PART 2

Vertebrates

ALTRUISM, INCLUSIVE FITNESS, AND EVOLUTIONARY GAME THEORY

M. Treisman

Department of Experimental Psychology
University of Oxford
England

INTRODUCTION

In this paper I want to develop two methods for warning behavior, consider their implications for the concept of inclusive fitness, and derive some consequences. The occurrence of altruism is a central problem in understanding social behavior. The term is applied to behavior which favors the evolutionary success of a recipient member of a species at some cost to the actor. It may but need not involve a social interaction. The selective effects of such behavior will depend on the genotypes of the conspecifics affected and so its selection is frequency-dependent. Three main approaches have been employed to analyse the evolution of altruism. These are based on the concept of "inclusive fitness", population genetic analysis, or evolutionary game theory.

Evolutionary analysis has traditionally relied on the concept of fitness, usually defined in terms of the number of viable offspring produced. This is a crude measure. Evolution proceeds by statistical changes in the gene pool, and an exact account must proceed by analysing these changes. The concept of Darwinian fitness is convenient because in many cases this will be linearly related to the expected effect on the gene pool. These are cases in which individual advantage does not involve interactions, competitive or cooperative, between different members of a species, and the effect of an allele is similar for each animal possessing it.

But when we consider social behaviors or other interactions we must draw implications about the fate of an allele from the experiences of individuals possessing that allele whose fates may differ as a result of interactions with each other. It is difficult to

91

apply the classical approach to behavior which has different effects
on interacting animals who are genetically the same. An individual
altruist who performs a benevolent act may himself suffer a loss as
a result. It is difficult to define his fitness so as to correct
for these complexities, and it might seem best to attempt to cal-
culate gene pool changes directly. But in practice considerable
thought has been given to finding corrected measures of fitness,
such as "inclusive fitness", which could be used to anticipate the
course of selection in these complex cases.

INCLUSIVE FITNESS

Hamilton (1963, 1964a, 1964b) conjectured that altruism would
be selected when the absolute value of the ratio of the benefit to
the recipient to the cost to the donor exceeded a criterion for
which I shall use the symbol k_H. Suppose that an altruist is a
member of a group of N kin, of average relatedness r, and that he
confers on each other member of the group a benefit A_b, at a cost
to himself given by A_c, these quantities being considered as increments to fitness. Then altruism will be selected provided that

$$k = \left| A_b \; A_c \right| > \frac{1}{r(N-1)} = k_H \tag{1}$$

We may also write this as the claim that the net benefit must
be greater than zero (West Eberhard, 1975):

$$A_c + r(N-1)A_b > 0, \tag{2}$$

for selection for altruism to proceed. This equation defines the
"inclusive fitness" which is substituted for classical individual
fitness.

I shall use the term a "Hamilton criterion" to mean any constant threshold quantity which must be exceeded by the benefit-
cost ratio for selection to occur. Such quantities may differ in
definition from k_H so long as they are not frequency-dependent
If a "Hamilton criterion" is exceeded selection will proceed to
fixation.

There have been a number of attempts to put Hamilton's (1963)
proposal on a rigorous basis. Some have succeeded in deriving
Equations 1 or 2, others have obtained different results (Treisman,
in press). What determines whether a model will support the inclusive
fitness criterion or not? One possibility is that this depends on
the assumptions made about the way contributions to fitness combine
together.

Genetics has been well studied and the appropriate assumptions
are usually clear. But since we know far less about how a series

of events may interact to determine the probability of replication than we do about the genetics of that replication, assumptions about fitness are often chosen arbitrarily, to give a convenient form, and otherwise do not receive much attention. But examination of the literature suggests the possibility that, at least on some genetic assumptions, linear fitness combination rules will ensure positive selection once the cost-benefit ratio is above a Hamilton criterion, nonlinear fitness rules will not give this simple relation. I shall examine this by comparing models for altruism which may employ different rules.

MODELS FOR ALTRUISTIC WARNING

Two combinatorial models for altruistic warning will be presented. We start by considering how we should combine the effects on fitness of a succession of events, in a simple model of altruism. A haploid population is subject to predation during a period of maturation prior to mating. An altruist who observes a predator may warn his fellows. During the growth period predators make $\underline{t}$ attacks, and each animal has a similar probability, $\underline{p}$, of surviving an unsignalled attack. If animal $\underline{i}$ survives to mate it will contribute $\underline{G_i}$ successful gametes to the gene pool. So its expected fitness is $\underline{F_i} = \underline{p}^t\underline{G_i}$.

But some animals may benefit on one or more occasions from hearing an altruist's warning. If $\underline{i}$ receives $\underline{w}$ warnings, and each warning increases its probability of escaping from that attack from $\underline{p}$ to $\underline{p_1}$ then its expected contribution to the next gene pool becomes

$$F_i(w) = p_1^w p^{t-w} G_i = (p_1/p)^w p^t G_i = B^w F_i \tag{3}$$

where $\underline{B} = \underline{p_1}/\underline{p} > 1$. So a warning multiplies the hearer's fitness by a constant, $\underline{B}$, whatever stage in the sequence it comes. It also imposes a cost on the altruist: his cry lowers his own survival probability to $\underline{p_0} < \underline{p}$. Thus his prior fitness is multiplied by $(\underline{p_0}/\underline{p}) = \underline{C} < 1$.

An alternative would be to express the effect of the warning as a fitness increment. But then the value of this increment depends on when the warning occurs and on the events on other occasions (Treisman, in press). For this reason we will prefer a multiplicative rule for taking account of an altruistic act. The first model to be described is referred to as the "concealment" model.

The Concealment Model

We assume a large, panmictic, discrete generation, haploid

sexual population of initial size U and consisting of two genotypes, L, which supports altruistic behavior, and l, which does not.
A proportion r of the animals are of genotype L. The population is
otherwise stable: in the absence of altruistic behavior the representation of each type would be multiplied by unity to give the
next generation. Predators make occasional attacks during a growing
season preceding mating.

Consider an extreme case, in which there is only one altruistic
response to an attack by a predator in each generation. A single
group of N animals has formed and includes at least one altruist.
A predator attacks: an altruist utters a warning. What effect will
this have on the representation of L in the gene pool of the next
generation?

The group consists of kin, the probability that any two members
possess alleles identical by descent being r. The alarm-raiser must
possess L. The genetical similarity, r_s, of the donor to any other
member of the group will depend on r and on the prevalence of L in
the population, $\bar{r}$:

$$r_s = r + (1-r)\bar{r} .\tag{4}$$

Then the expected number of members of the group who are of
genotype L is $1 + r_s(N-1)$. The remainder of the population has n
replicas of L, so that the initial proportionate representation of
L in the population is

$$R_o(L) = \frac{1 + r_s(N-1) + n}{U}\tag{5}$$

We may also take this as the genetic similarity at this locus
of the alarm-raiser to the (large) population, i.e. $\bar{r} = R_0(L)$. The
cost of uttering the alarm call causes the altruist's expected contribution to the next generation to be multiplied by $C < 1$. This
cost may also be written as $A_c = C - 1 < 0$. The other members of
the group benefit: the alarm call multiplies their expected replication by $B > 1$ in each case. We can also write $A_b = B - 1 > 0$.

If there had been no cry the prior fitness would be one in all
cases. But now benefits have accrued to members of both types, L
and l. So the first filial generation will have

$$U_1 = U - N + C + B(N-1)\tag{6}$$

members, of which the expected number of L genotypes will be

$$E[N_1(L)] = C + Br_s(N-1) + n$$

and so the proportionate representation of $\underline{L}$ will become

$$R_1(L) = E\left[N_1(L)\right] / U_1. \tag{7}$$

Selection for $\underline{L}$ requires a positive change in its expected share of $\underline{U}_1$:

$$\Delta U(L) = U_1\left[R_1(L) - \bar{r}\right] > 0 \tag{8}$$

which can also be written

$$\Delta U(L) = (1-\bar{r})A_c + r(1-\bar{r})(N-1)A_b > 0. \tag{9}$$

Equation 9 corresponds to the inclusive fitness equation (2) and will give Hamilton's (1963, 1964a, 1964b) criterion. Although this supports inclusive fitness the situation is unrealistic: limiting the number of altruistic acts to one necessarily linearises the relation between the number of acts and the fitness change resulting.

We now consider a more complete version of the model in which the $\underline{U}$ animals in the initial generation are divided into n_g groups each of size $\underline{N}$. Every group can expect to be attacked once by a predator, during the period prior to mating. Fot protection against this attack the animals rely mainly on concealment. Other things equal, each animal is normally sufficiently well hidden for the chance that the predator will fail to detect and capture him to be $\underline{k} < 1$. Then his resultant fitness (in the absence of altruists) is given by multiplying his prior fitness by $\underline{k}$. It is convenient to assume that, leaving altruism aside, the population would be stable, in which case we may set the prior fitness to $(1/\underline{k})$. Thus the resultant fitness of an unwarned non-altruist is 1.

Animals with the allele $\underline{L}$ act as altruists by taking on the role of sentinel. This involves that they add two items of behavior to their repertoire. First, they keep watch for the predator. Second, if an altruist detects him, and there has been no prior warning, he utters an alarm cry. A cost will arise for the caller, as he may attract the attention of the predator toward himself, thus reducing his chance of going undetected to $\underline{C}k$, with $\underline{C} < 1$. Other animals, hearing the alarm cry, will take steps to conceal themselves more effectively and so increase their chances from $\underline{k}$ to $\underline{B}k$, with $\underline{B} > 1$. If the predator is not detected, the fitness of an altruist will be the same as that of a non-altruist. Non-altruists do not attempt to detect the predator. Will altruism be selected, and if so, under what conditions?

The mean representation of $\underline{L}$ in the initial population is $\bar{r} = R_0(L)$. The animals in a group are recruited randomly with regard to the possession of L, save that there may be a degree of relatedness, r, $0 \ll r \ll 1$, between members of a group. For convenience, we may suppose this derives from a multiple insemination population structure: the members of a group share one common ansector, for example, they may all be hatched from the eggs of a common mother, and each member's second ancestor is chosen randomly and independently from the population at large. Let the probability that a given individual has an allele identical by descent with that of the common ancestor be $r^{\frac{1}{2}}$. Then the probability that two randomly hosen members of a group both possess replicas by descent of the common ancestor's allele is r. They may also be genetically the same though not by descent if both have the same allele but it comes in at least one case from the randomly chosen ancestor. So with probability $r^{\frac{1}{2}}$ a member carries a replica of the common ancestor's allele, and with probability $(1 - r^{\frac{1}{2}})$ his genotype is sampled randomly from the population.

The members of each group are outcomes of the random mating of the generation before, so that the constitution of a group may be described by the binomial distribution. If the allele of the common ancestor is represented by λ (and such ancestors are a random sample), then the probability over all groups that this allele is L must be $p(\lambda = L) = \bar{r}$.

For any given group, λ was either L or it was l. In the first case, the probability that any individual member of the group has L is $r_1 = r^{\frac{1}{2}} + (1 - r^{\frac{1}{2}})\bar{r}$ and in the second case it is $r_2 = (1 - r^{\frac{1}{2}})\bar{r}$. Thus $\bar{r}$ determines both r_1 and r_2, and the relative proportions of the two types of groups.

Each group is placed once under threat by a predator. If there is more than one altruist in a group, more than one individual may detect the predator. If each altruist has the same probability of detecting the attacker, q, $0 \ll q \ll 1$, the simplest possible assumption about the altruists' joint behavior is that they are independent detectors. Then if a group has x altruists the probability that a warning cry will be uttered is

$$P(D|x) = 1 - (1-q)^x. \qquad (10)$$

The first altruist to discover the predator will arouse the other members of the group, who thereupon take action. Once a warning has been uttered, there is no advantage in further calls, so we assume they are not made. Thus, the alarm is given once only at most, and we are concerned whether or not at least one altruist sees the predator.

Some groups receive a warning when the predator attacks, others do not, and their potential reproduction varies accordingly. So the size of the gene pool in the first filial generation will be

$$U_1 = (1/k)P(\lambda{=}L)n_g\sum_{x=0}^{N} \begin{bmatrix}N\\x\end{bmatrix} r_1{}^x(1{-}r_1)^{N-x}\bigl[P(D|x)\ (C{+}(N{-}1)B)k +$$

$$(1{-}P(D|x))Nk\bigr] + (1/k)P(\lambda{=}1)n_g\sum_{x=0}^{N}\begin{bmatrix}N\\x\end{bmatrix} r_2{}^x(1{-}r_2)^{N-x}\bigl[P(D|x)(C +$$

$$(N{-}1)B)k + (1{-}P(D|x))Nk\bigr]. \tag{11}$$

From this we can determine the proportionate representation of
L in the population, $R_1(L)$ (Treisman, in press). The requirement for
selection for L (Equation 8) becomes

$$\Delta U(L) = k_c(r,\bar{r},q,N)n_gA_c + k_b(r,\bar{r},q,N)n_gA_g > 0 \tag{12}$$

where k_c and k_b are each complicated functions of r, $\bar{r}$, q and N. We
may rewrite this equation as $A_c + [k_b/k_c]A_b > 0$.

Provided $k_b > 0$, this can be expressed as the requirement that

$$\left|\frac{A_b}{A_c}\right| > \frac{k_c}{k_b} = k(L). \tag{13}$$

Here $k(L)$ is a criterion that the benefit-cost ratio must exceed
for selection for L to proceed. But this criterion is not a Hamilton
criterion as both $\bar{k}_b$ and k_c are frequency-dependent. This is not
compatible with and could not be reduced to the inclusive fitness
equation (2). We have a more complex relation than is assumed by
that equation. Although measures of fitness can be derived from the
present analysis (Treisman, in press) they are not simple and not
necessarily helpful in the analysis of the situation.

Figure 1 illustrates some implications of the model by plotting
the criterion $k(L)$. Like k_H, this is a function of r and N. But it
also varies with $\bar{r}$, the prevalence of L, and with q, the efficiency
of detection.

We may note the following points: (1) If a positive value of
$k(L)$ exists, it is greater than Hamilton's inclusive fitness
criterion, k_H. Use of the inclusive fitness criterion instead of
$k(L)$ would lead us to over-estimate the likelihood of selection
for altruism. (2) The figure shows that the requirement $k(L)$ becomes
more stringent the less the relatedness, r, within the group; (3)
the more altruists there are already in the population ($\bar{r}$); and (4)
the more efficient they are (q). Thus as the likelihood of receiving
a warning increases, resistance to further selection for L also
increases. (5) The right lower panel (in which log scales are used)
illustrates that selection for L is less likely in larger groups.
Here k_H is given by a straight line of slope -1 for a fixed group

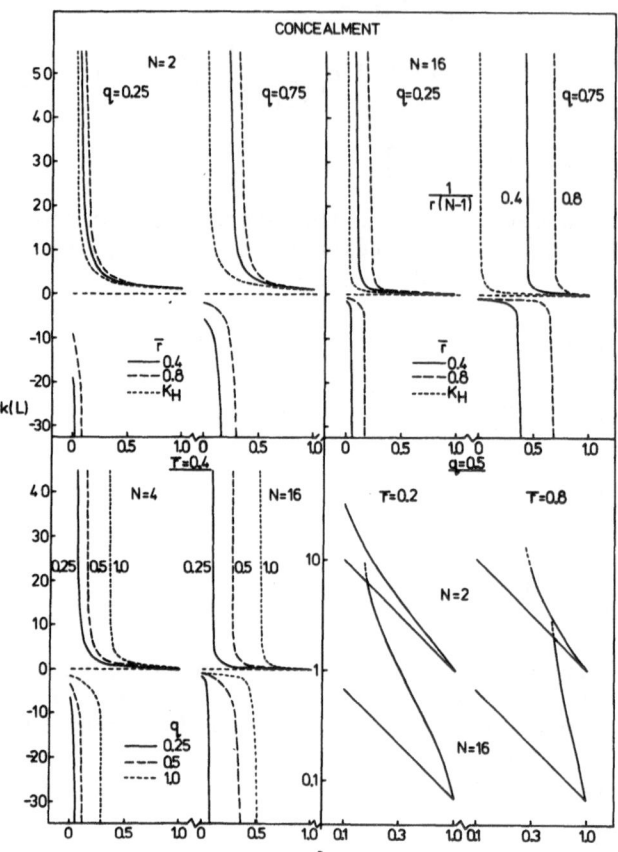

Fig. 1. The Concealment Model. If k(L) is negative then selection
        for altruism cannot occur. If it is possitive it requires
        to be exceeded by |A_b/A_c|. In the upper panels k(L) is
        plotted against r, the probability of identity by descent,
        for two levels of the efficiency of detection by altruists,
        q=0.25 and q=0.75, and for two levels of L in the population
        as a whole: r̄=0.4 and r̄=0.8, for groups of size N=2 or N=16.
        The value of Hamilton's criterion, k_H, is given by the
        dotted line. The lower left panel plots k(L) against r for
        r̄=0.4, for N=4 or 16, and for q=0.25, 0.5 or 1. The right
        lower panel presents a log-log plot of positive values of
        k(L) against r, for r̄=0.2 (on the left) or 0.8 (on the
        right), N=2 (above) or 17 (below) and q=0.5. Here the
        curved lines show k(L) and the straight lines the corre-
        sponding values of k_H.

size, since log k_H = -log r - log(N-1). k(L) may considerably exceed k_H, but converges toward k_H when r is high. The excess is greater (where log k(L) exists) for the larger group than for the smaller.

In short, as the number of altruists in the population rises, selection for L becomes more difficult. It requires greater relative benefits for recipients than is given by Hamilton's criterion. It requires degrees of relatedness in the upper part of the range: there is no simple trade-off between number of beneficiaries and r. Selection also becomes more difficult if q increases and if N increases: any factor which raises the "effective density of altruism", the number and efficiency of altruists in the population, opposes further selection.

The Evasion Model

In a second type of altruistic warning model, all animals attempt to detect the predator and are equally likely to do so, but only altruists share their knowledge with the rest. A second difference from the Concealment model is that animals escape from the predator by flight. The probability that an unwarned animal will detect the predator in time to make its escape is q. If it does not detect the predator it is captured. Thus detection is essential for survival. This model is otherwise similar to the Concealment model.

A cost arises for the caller, given by the multiplier $C < 1$, as he will attract the attention of the predator toward himself. The benefit of the alarm cry to others may arise from a number of sources. First, $P(D|x)$ increases with x. So the more animals share information, the more likely is each to be aware of the predator. Second, the alarm cry may energize the hearers, and by attracting the attention of the predator to the caller facilitate the escape of the rest. Third, if the predator approaches over a period of time there will be an advantage in detecting him early rather than late. He is more likely to be detected sooner if the probability of detecting him is high. The effect of the increased probability of detection is represented by $P(D|x)$, and other constant beneficial effects of the cry are represented by the multiplier $B > 1$.

The expectation of survival of a solitary animal is given by his probability of detecting the predator, q. This will multiply his prior fitness to give the reference fitness level, 1 (since we assume that apart from the effects of shared detections the population would be stable). Thus we may set a prior fitness (or scaling factor) at (1/q), and multiply all returns from the encounter with the predator by this factor to give resultant fitness. At a given level of x, the expected return to an altruist who utters an alarm cry will be $P(D|x)C$ and to all other altruists and non-altruists it will be $P(D|x)B$. If no altruists makes a detection the return for altruists is 0, while that for each non-altruist in the group is (1 - $P(D|x)$)q.

This model may be analyzed in the same way as the Concealment model (Treisman, in press). The requirement for selection for $\underline{L}$ becomes

$$\Delta U(L) = k_c(r,\bar{r},q,N)(n_g/q)A_c + k_b(r,\bar{r},q,N)$$
$$(n_g/q)A_b + k_o(r,\bar{r},q,N)(n_g/q) > 0 \tag{14}$$

or

$$A_c + (k_b/k_c)A_b + (k_o/k_c) > 0, \tag{15}$$

where $\underline{k_o}$, $\underline{k_b}$ and $\underline{k_c}$ are all frequency-dependent. Again we have a result which is not compatible with the inclusive fitness criterion Equation 2.

Unfortunately, Equation 15 does not allow us to examine the implications of the model by calculating a critical benefit-cost ratio. But for a given value of $\underline{B}$ we can find the corresponding value of $\underline{C}$, $\underline{C_l}$, at which $\Delta U(L) = 0$. Selection for $\underline{L}$ will proceed only if $\underline{C} > \underline{C_l}$. $\underline{C_l}$ must, of course, be greater than or equal to 0. It can also be greater than 1. If so, this will tell us that altruism is not possible: a caller must receive a positive benefit before the behavior of giving the alarm will be selected. In Fig. 2, $\underline{C_l}$ is plotted against $\underline{r}$ for various parameter values. $\underline{C_H}$, the critical value of $\underline{C}$ corresponding to Hamilton's criterion, i.e. the value given by solving $(\underline{B} - 1)/(1 - \underline{C_H}) = \underline{k_H}$, is shown as a dotted line for each set of parameters.

Figure 2 illustrates a number of points. (1) In no case does selection for $\underline{L}$ take place if $\underline{r} = 0$ and $\underline{C} < 1$. (2) Selection will take place for any value of $\underline{C}$ provided $\underline{B} > 2$ (of the cases examined) and $\underline{r}$ is sufficiently high. That is, $\underline{C_l}$ always falls to 0 before $\underline{r}$ reaches 1. (3) $\underline{C_H}$ (the inclusive fitness value), usually, but not in all cases, over-estimates the likelihood of selection for $\underline{L}$. (4) Selection may become more difficult when $\underline{q}$ increases, and (5) when $\bar{\underline{r}}$ increases. (6) An increase in $\underline{B}$ may require a higher value of $\underline{C}$ (that is, a lesser cost of altruism) at low values of $\underline{r}$. (7) Selection is more difficult in large groups. (8) Finally, we note that in the present model selection may take place when $\underline{B} = 1$ ($A_b = 0$), something which would be impossible if the inclusive fitness equation applied. This can occur because of the benefit obtained by both donors and recipients from the increase of $\underline{P(D|x)}$ with $\underline{x}$. This common benefit means that the net effect on an altruist may be positive if $\underline{C}$ is not too small. The horizontal bars in Fig. 2 indicate the values of $\underline{C}$ below which there is a true net cost to the altruist (Treisman, in press). For larger values of $\underline{C}$ the behavior cannot be considered altruistic.

In general, at low values of $\underline{r}$ selection requires levels of $\underline{C}$ which do not constitute altruism. Only when $\underline{r}$ is high will giving the alarm be selected despite a negative effect on the altruist.

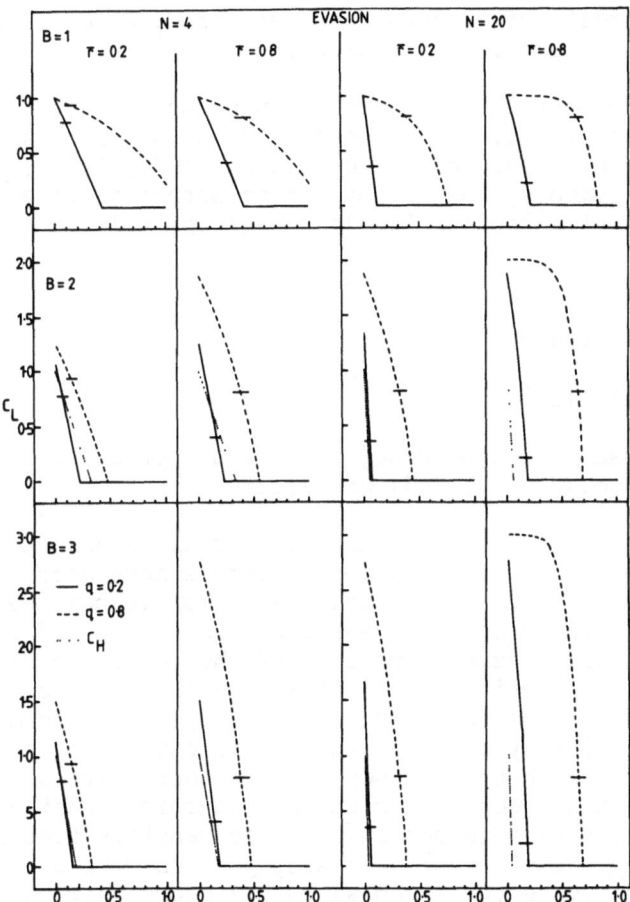

Fig. 2. The Evasion Model: Values of C_L, the level of C which must
be exceeded for selection for L to occur, are plotted
against r for N = 4 or 20, B = 1, 2 or 3, $\bar{r}$ = 0.2 or 0.8,
and q = 0.2 or 0.8. C_H (derived from Hamilton's criterion
k_H) is shown as a dotted line.

LINEAR FITNESS COMBINATION

   The present models employ nonlinear fitness assumptions: first,
because once an animal has been warned that an attack is imminent,
it does not gain an equal additional benefit from a second or third
warning. Second, because the probability of a warning increases not
in direct proportion to the number of altruists but as a function

of the probability of at least one detection, which is a nonlinear function of the number of altruists.

We must compare the results of these models with those that would be given by a linear fitness combination rule. A model of this kind has been put forward by Wilson (1975, 1977). He assumes that, in our symbols, the fitness of an altruist, as a function of the number of altruists ($\underline{x}$) in the group, will be

$$F(\text{Altruist}) = A_c + (x - 1)A_b$$

and of a non-altruist

$$F(\text{Non-altruist}) = xA_b.$$

These assumptions provide, as we would predict, a result which agrees with Hamilton's (1963) criterion.

The linear fitness rule supports and the nonlinear rule rejects the inclusive fitness formulation. Which is more appropriate for warning behavior in the real world? A rationale has been given for the nonlinear assumptions we have preferred. It is of interest to consider what would have to be true of the world for Wilson's assumption that benefits are directly proportional to the number of altruists to be valid. There are two ways in which his expressions could arise: (1) They could represent the outcome of a single attack by a predator. In that case they require that every altruist in the group detects the predator and utters a warning: that every altruist suffers the cost of this action but also benefits from every other warning cry ($\underline{x}$ - 1 of them) he hears, just as does a non-altruist. Further, each cry he (or a non-altruist) hears carries the same additional advantage, however many have already been heard. Four cries heard will double the safety of a hearer as compared with two, and two will double his safety as compared with one. (2) Alternatively, the expressions could represent the accumulated benefits over $\underline{x}$ attacks. An altruist would raise his voice on one of these attacks at the cost A_c and benefit on the other $\underline{x}$ - 1. A non-altruist would benefit on all. But this requires that the number of attacks on a group must equal the number of altruists in it. There is no mechanism which would bring this about. It is no solution to suppose that there are at least $\underline{N}$ attacks on every group (we need $\underline{N}$ since some groups will consist wholly of altruists) because we must then explain why altruists fall silent after exactly $\underline{x}$ such attacks, and sit out the rest. These accounts do not match our common experience of the world.

DISCUSSION

Our comparison of the models leads to the following conclusions:

1. It is important to distinguish between linear and non-linear fitness combination rules. The former lead to Hamilton's criterion, the latter may not. But, for the situations we have considered, non-linear rules give a more reasonable account of biological reality. This raises serious problems for the unquestioning application of "inclusive fitness" arguments.

2. Models of warning behavior must take account of possible general benefits from information sharing, as well as the individual costs and benefits. In the Evasion but not the Concealment model, the common benefit of information sharing in detection offset the cost of uttering an alarm cry.

3. High values of r are an important precondition for selection of the behavior. This is demonstrated by the requirement of values of $C > 1$ for some parameters and r low in the Evasion model: when r is low the benefit of a call goes mainly to non-bearers of the allele L. Thus selection when $C < 1$ requires either that r is high, or that kin and non-kin can be distinguished and benefits directed to the former and not the latter.

The results for both models suggest that, except in clones, selection may not proceed so far as to make altruism universal. We are led to the question whether an evolutionarily stable level of altruism can be defined for warning behavior, and if so how to specify it. I describe below some extensions to the models which provide answers to this question.

EVOLUTIONARY GAME THEORY

Evolutionary game theory has been developed for analysing the evolutionary control of aggression in haploid species by Maynard Smith and his colleagues (Maynard Smith, 1974, 1978; Maynard Smith and Price, 1973; Maynard Smith and Parker, 1976). However, exact results can be obtained for diploid species (Treisman, 1981) and the approach can be applied to altruism (Treisman, 1982b). I have developed game theory accounts of both Concealment and Evasion in dyads, and some results are illustrated in Fig. 3.

In all cases, these models predict that there will be no altruism if $r = 0$. The results in Fig. 3 apply for $r = \frac{1}{2}$, that is for siblings in the diploid species. The probability of choosing the altruistic option at the ESS is represented by $p_I(r)$, for haploid species, where r is the relatedness between contestants, and by $d_I(r)$ for diploid species. For diploid species the corresponding stable polymorphism, the proportion of the altruistic allele in the evolutionarily stable population, is given by $w(r)$. Haploid species in which individuals follow one or other pure strategy are described as heterogeneous, and those whose members apply the mixed strategy specified by the ESS are homogeneous.

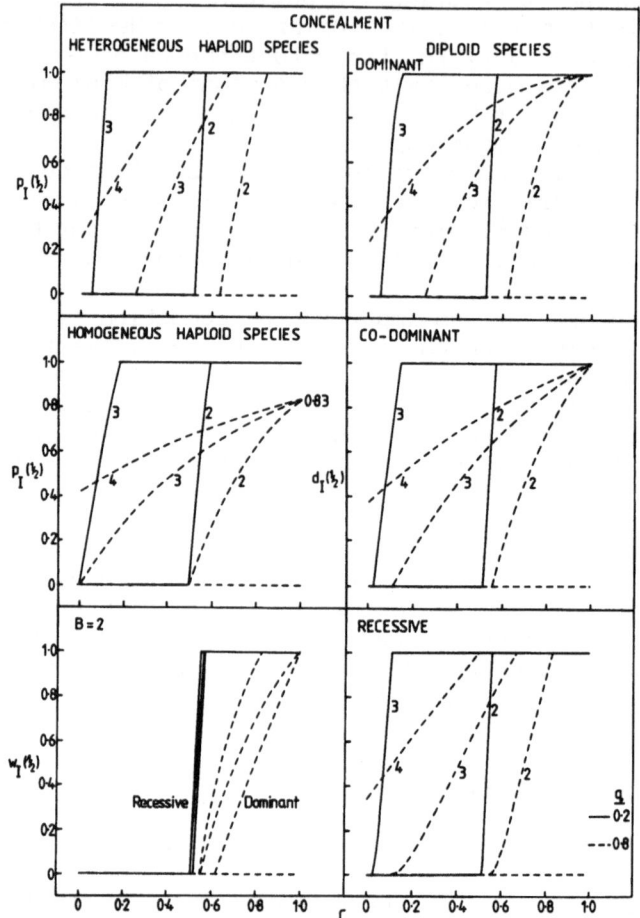

Fig. 3. Evolutionarily Stable Strategies and polymorphisms for the game "Concealment" for r=0.5: the two upper panels on the left show values of $p_I(\frac{1}{2})$, the probability of choosing the altruistic strategy given by the ESS for heterogeneous haploid species (upper panel) and homogeneous haploid species (middle panel) for B=2, 3 and 4 (shown as parameters on the curves) plotted against C. On the right, $d_I(\frac{1}{2})$, the evolutionarily stable probability of warning behavior for diploid species, is shown for dominant inheritance, co-dominant inheritance, and recessive inheritance. The left bottom panel shows the corresponding stable polymorphisms $w_I(\frac{1}{2})$ for diploid species for B=2. These curves are Recessive, Co-dominant and Dominant going from left to right. Curves are shown for q=0.2 (continuous lines) and q=0.8 (dashed lines). No curve is shown for B=4 and q=0.2 as $p_I(\frac{1}{2})$ and $d_I(\frac{1}{2})$ 1 in all cases for these parameters.

While there are differences between the different types of inheritance certain findings are common to them all.

1. The probability of warning behavior increases as C increases.
2. It increases as B (the parameter on the curves) increases.
3. It increases as its effectiveness (q) falls.
4. When q is high the evolutionarily stable strategies take on intermediate values over a range of levels of C. This is a result incompatible with the concept of "inclusive fitness". If the latter is sufficient for selection of altruism, it should take such selection to fixation.
5. But when q is low, $p_I(\frac{1}{2})$ and $d_I(\frac{1}{2})$ shift sharply from 0 to 1 over a small range, approximating the step function that would be produced by a Hamilton criterion. This may be attributable to the fact that $p(D|x)$ is approximately linear with x when both q and x are small but not for q large. For "Concealment" the value of C at which the nearstep occurs corresponds to a benefit-cost ratio of approximately 2.2, near the value of Hamilton's criterion for $r = \frac{1}{2}$. But this is simply a function of the parameter q, as can be seen by comparing the curves for $q = 0.2$ and $q = 0.8$.
6. Similar results are obtained for "Evasion" except that the curves are all shifted to the left, the shift being more marked when q is low. Thus the benefit-cost ratio at which the "Evasion" $p_I(\frac{1}{2})$ or $d_I(\frac{1}{2})$ shifts from 0 to 1 for $B = 2$, $q = 0.2$, is approximately 1.1. The greater frequency of warning behavior with Evasion reflects our earlier observation that in this case the enhanced probability of detection partially outweighs the cost to the altruist.

It appears that if individuals are efficient at detecting a predator, the probability of altruism may be at an intermediate level for a wide range of costs. But if q is low the picture shifts toward the "biological flip-flop" constituted by a threshold criterion. At one value of C no bird will warn. At a slightly higher value warning will become universal.

COMBINATORIAL MODELS IN DIPLOID SPECIES

The game theory analysis of the behavior of pairs of animals is a special case of the analysis of the effects of altruistic behavior on gene frequency changes. The models of warning altruism for haploid groups presented earlier make the same assumptions as the game theory models. It can be shown that the combinatorial models reduce to the game theory models when $N = 2$. Similar combinatorial models have been developed for diploid sexual populations (Treisman, 1982b), and some results for "Concealment" are illustrated in Fig. 4.

The figure illustrates a number of points.
1. While under some conditions there may be no altruism, and under others it may become universal, there are ranges of C over which stable polymorphisms exist.

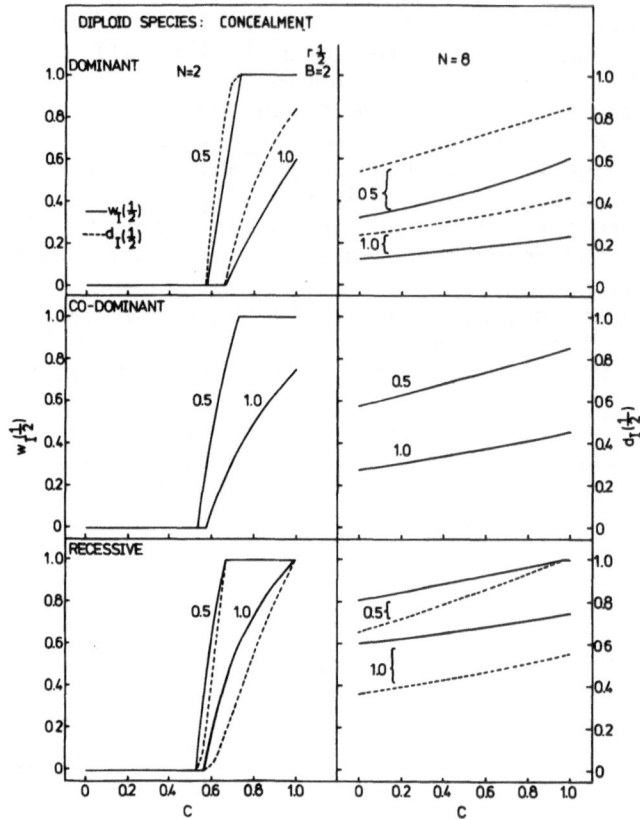

Fig. 4. Concealment Model for siblings ($r = \frac{1}{2}$) in diploid populations. The stable polymorphism, $\underline{w}_I(\frac{1}{2})$, is shown as a continuous line and the corresponding ESS ($d_I(\frac{1}{2})$) as a dashed line for dominant and recessive diploid inheritance of altruism for group sizes 2 and 8, $\underline{B} = 2$, $0 < \underline{C} < 1$, and q = 0.5 or 1 (shown as the parameters on the curves). In the co-dominant case, $\underline{d}_I(\frac{1}{2})$ is identical with $\underline{w}_I(\frac{1}{2})$.

2. Altruism is much less sensitive to the value of $\underline{C}$ in large groups than in small. In large groups it occurs at lower levels of $\underline{C}$, but does not become as dominant as in small groups when the cost is small. In small groups it may approximate a step function.

3. A reduction in the probability of detection ($\underline{q}$) gives an increase in the stable level of altruism. It may seem paradoxical that the prevalence of altruism rises when its efficiency falls. But it is when $\underline{q}$ is low that $\underline{P}(D|\underline{x})$ increases most markedly with $\underline{x}$.

4. The stable strategies (dashed lines) vary much less with the degree of dominance than do the stable polymorphisms (continuous lines). $\underline{w}_I(\frac{1}{2})$ adjusts itself to achieve a similar phenotypical result for each genetic system.

5. Similar conclusions apply to the Evasion model, except that warning behavior is more frequent for all parameters.

If we set $\underline{B}$ = 1 we do not expect stable altruism with the Concealment model. But it may be shown for intermediate values of $\underline{q}$, in groups which apply Evasion. For $\underline{N}$ = 8 it is less evident at low values of $\underline{C}$ than when $\underline{B}$ is larger. But as $\underline{C}$ approaches 1.0 the curves may rise steeply and exceed those for $\underline{B}$ = $\overline{2}$. It seems that when $\underline{B}$ is small, sensitivity to $\underline{C}$ increases. These effects are shown in Fig. 5.

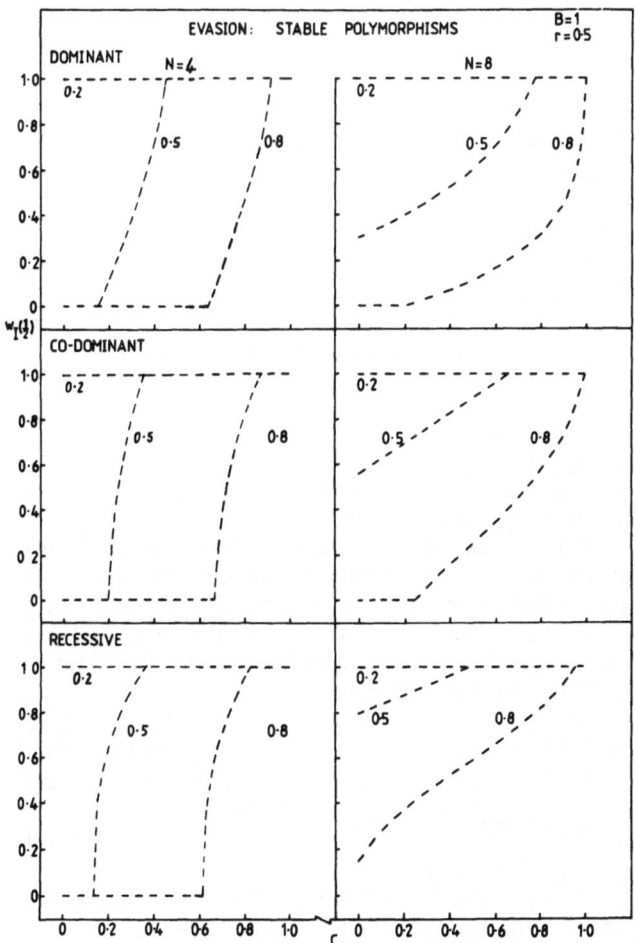

Fig. 5. Evasion Model. The stable polymorphism $\underline{w}_I(\tfrac{1}{2})$ is shown for $\underline{B}$ = 1.0 and $\underline{q}$ = 0.2, 0.5 or 0.8, in groups of size 4 and 8.

CONCLUSIONS

The arguments presented here illustrate that "inclusive fitness" is limited in its application, and does not provide a useful description of classes of behavior in which monlinear processes may occur.

Some results of the present models may be of general interest. Of special note are the absence of altruism when $r = 0$ and the relative resistance to selection for altruism at low levels of relatedness, the greater prevalence of altruists when the efficiency of detecting a predator, is low, the reduced sensitivity to the cost of altruism in large groups, the stronger selection in species relying on evasion than on concealment, and the importance of distinguishing between altruism and common reciprocal benefits, both of which may be present in a given situation.

REFERENCES

Hamilton, W. D., 1963, The evolution of altruistic behavior, Am. Nat., 97:354-356.

Hamilton, W. D., 1964a, The genetical evolution of social behavior, I, J. Theor. Biol., 7:1-16.

Hamilton, W. D., 1964b, The genetical evolution of social behavior, II, J. Theor. Biol., 7:17-51.

Maynard Smith, J., 1974, The theory of games and the evolution of animal conflicts, J. Theor. Biol., 47:209-221.

Maynard Smith, J., 1978, Optimization theory in evolution, Ann. Rev. Ecol. Syst., 9:31-56.

Maynard Smith, J., and Parker, G. A., 1976, The logic of asymmetric contests, An. Behav., 24:159-175.

Maynard Smith, J., and Price, G. R., 1973, The logic of animal conflict, Nature, 246:15-18.

Treisman, M., 1981, Evolutionary limits to the frequency of aggression between related or unrelated conspecifics in diploid species with simple Mendelian inheritance, J. Theor. Biol., 93:97-124.

Treisman, M., 1982a, Models for the evolution of altruistic warning in a haploid species, with implications for the concept of inclusive fitness (in press).

Treisman, M., 1982b, Evolutionary stable levels of altruism in haploid and diploid species, J. Theor. Biol., 97:437-480.

West Eberhard, M. J., 1975, The evolution of social behavior by kin selection, Quart. Rev. Biol., 50:1-33.

Wilson, D. S., 1975, A theory of group selection, Proc. Nat. Acad. Sci. (U.S.A.), 72: 143-146.

Wilson, D. S., 1977, Structured demes and the evolution of group advantageous traits, Am. Nat., 111:157-185.

ANTIPREDATOR ADAPTATIONS OF SALAMANDERS: EVOLUTION AND CONVERGENCE AMONG TERRESTRIAL SPECIES

Edmund D. Brodie, Jr.

Biology Department, Adelphi University
Garden City
New York 11530, U.S.A.

INTRODUCTION

"The exceptional standard of defence has been only reached through the pressure of an exceptional need" (Poulton, 1890). Poulton realized that any new defensive mechanism acquired by a prey species would be followed by the evolution of more effective predatory mechanisms or increased resistance by its predators. A result of these ever present selective pressures on both predator and prey is the evolution of intricate and efficient antipredator mechanisms.

Terrestrial salamanders have evolved a remarkable diversity of mechanisms that help them avoid being eaten by predators (Table 1). These mechanisms have been enumerated and discussed elsewhere (Brodie, 1977; Brodie et al., in press; Dodd and Brodie, 1976). Experimental studies have demonstrated that the following proposed antipredator mechanisms enhance the survival of salamanders: distasteful skin secretion (Brandon et al., 1979a, 1979b; Brodie et al., 1979; Brodie and Formanowicz, 1981; DiGiovanni and Brodie, 1981; Dodd et al., 1974; Hensel and Brodie, 1981; Hubbard, 1903); aposematic (warning) color patterns (Hensel and Brodie, 1976; Johnson and Brodie, 1975); pseudoaposematic (mimetic) color patterns (Brodie and Brodie, 1980; Brodie and Howard, 1973; Howard and Brodie, 1971, 1973); immobility (Dodd and Brodie, 1976; Ducey and Brodie, in press); various behavior patterns (Ducey and Brodie, in press; Johnson and Brodie, 1975); biting (Brodie, 1978); and tail autotomy (Ducey and Brodie, in press).

Further support for the view that the characteristics listed in Table 1 are adaptive is the prevalence of convergence between members of different families. If two species belonging to different families have the same or very similar characteristics (e.g., color pattern, glandular distribution, or behavior patterns) the two alternative possibilities are (1) that the common ancestor of the two families also had the characteristic in question or (2) that the common ancestor did not have the characteristic and it has evolved independently in the two species. When two species evolve the same or very similar characteristics independently (convergence) it is likely that they responded to similar environmental pressures. In the case of antipredator mechanisms the environmental pressure would be predation.

Many of the characteristics listed in Table 1 act synergistically to enhance survival (Arnold, 1982; Brodie, 1977; Ducey and Brodie, in press; Hubbard, 1903), and specific behavior patterns are correlated with different morphologies.

The purpose of this paper is to point out the most common synergistic interactions between behavioral and morphological antipredator mechanisms and to examine the evolution of these sets of adaptations.

OPERATIONAL DEFINITIONS

Table 1 summarizes literature records and my own research on the antipredator mechanisms of terrestrial salamanders. A plus sign indicates that a character is present in at least one species of the genus. A minus sign indicates that a character is not present in any member of the genus that has been studied (this does not rule out the possibility that some species as yet unstudied may have the character). A blank space indicates that the information is not available. There is considerable variability among species of certain genera (notably *Triturus* and *Ambystoma*) in which some species exhibit certain characters not present in their congeners. Although many papers have been published on salamander antipredator adaptations no attempt has been made to establish operational definitions.

Following is an outline of the criteria used in determining the presence of each antipredator adaptation listed in Table 1. Adaptations h, l through w, y, and z are observed in response to actual or simulated predator attack. Predator attack may be simulated by pinching with forceps or tapping lightly with a probe (Brodie, 1977).

Noxious Skin Secretions

These repel at least some predators. It is common for avian and mammalian predators to be repelled by a salamander that is eaten readily by some snakes. In order for a genus to be listed as positive, at least one species has been observed to repulse a

predator under experimental conditions or the skin secretions burn
the mouth lining of humans. It is assumed, for the purpose of this
study, that secretions that cause pain to the mouth lining of humans
would have a similar effect on other mammals. Of the 29 genera listed as noxious 23 have been tested with predators; each is also re-
pulsive to humans. Both genera listed as negative are edible to a
wide array of predators and the skin secretions of those species
tasted do not burn the mouth of humans. All species of these two
genera have not been tested, but known intrageneric differences in
noxiousness are limited to levels of noxiousness, not presence or
absence. Most of the genera that have not yet been tested are probably noxious.

Toxic Skin Secretions

These have the potential to kill at least some types of predators
when ingested. Secretions that kill when injected but not when ingested are not considered to be "toxic". This usage recognizes as
toxic only those secretions which are toxic in the ecological sense.
Skin toxins are restricted to the Salamandridae and all species of
that family that have been tested are toxic. It is expected that
most, if not all, members of this family are toxic. All toxic
salamanders also have skin secretions that are very noxious to
predators.

Parotoid Glands

These are concentrations of enlarged granular glands located
in the skin at the rear corners of the head. These range in size
from very large obvious swellings as in *Salamandra* or some *Ambystoma*
to small concentrations as in *Ensatina*. Many literature records of
"parotoid glands" refer to the displacement of the skin in the
parotoid region by the epibranchials and ceratohyals and do not
actually refer to regions with increased sizes or numbers of granular glands.

Glandular Lateral Warts

These are concentrations of enlarged granular glands located
along the lateral or dorsolateral surfaces of the body. These concentrations may consist of only a few moderately enlarged glands
as in *Paramesotriton* or may consist of very large tightly packed
glands as in *Salamandra*. The "lateral warts" reported in the
literature for *Pleurodeles waltl* are actually thin aglandular
regions of skin (Eber, 1954; Nowak and Brodie, 1978).

Glandular Tail Dorsum

This refers to concentrations of enlarged granular glands
along the dorsal ridge of the tail.

Table 1. Antipredator adaptations of terrestrial salamanders. See text for criteria.

	a	b	c	d	e	f	g	h	i	j	k	l	m	n	o	p	q	r	s	t	u	v	w	x	y	z	aa	bb	cc
SALAMANDRIDAE																													
Tylototriton	+	+	+	+	+	+	+	+	−	+	+	+	−	−	+	−	+	+	+	+	+	+	−	−	−	−	−	+	+
Echinotriton	+	+	+	−	+	+	+	+	−	+	+	+	−	−	+	−	+	+	−	+	+	+	+	+	−	−	+	+	+
Pleurodeles	+	+	+	−	−	−	+	−	−	−	−	−	−	−	−	−	−	−	+	+	+	+	+	+	−	−	+	+	−
Salamandrina	+	+	+	+	−	+	+	−	−	−	+	+	−	−	−	−	−	+	+	+	+	+	−	−	+	−	+	+	+
Salamandra	+	+	+	+	+	+	+	+	+	+	+	+	+	+	+	+	+	−	−	−	−	−	−	−	−	−	−	+	−
Mertensiella	+	−	+	+	+	−	+	+	+	+	+	+	+	+	+	+	+	+	−	−	−	−	−	−	−	−	−	+	−
Chioglossa	+	−	−	+	−	+	+	+	+	+	−	−	+	−	−	−	−	−	−	−	−	−	−	−	+	+	−	−	−
Hypselotriton	−	−	−	−	−	−	−	−	−	−	−	−	−	−	−	−	−	−	−	−	−	−	−	−	−	−	−	+	−
Cynops	+	+	+	+	−	−	+	−	−	−	+	+	+	+	+	−	−	−	−	−	−	+	−	+	−	−	+	+	+
Notophthalmus	+	−	−	−	−	−	+	−	−	−	+	+	+	+	+	−	−	−	−	−	−	+	−	+	−	−	+	+	+
Taricha	+	−	−	−	−	−	+	−	−	−	+	+	+	+	+	−	−	−	−	−	−	−	−	+	−	−	+	+	+
Triturus	+	+	−	+	−	−	+	−	−	−	+	+	+	+	+	−	−	−	−	+	+	+	−	−	−	−	+	+	+
Paramesotriton	+	+	+	+	+	−	+	−	−	−	+	+	+	+	+	+	−	−	−	−	−	+	−	−	−	−	+	+	+
Neurergus	+	−	−	−	−	−	−	−	−	−	−	−	−	+	−	−	−	−	−	−	−	−	−	−	−	−	−	−	−
Euproctus	+	−	−	−	−	−	+	−	−	−	−	−	−	−	−	−	−	−	−	−	−	−	−	−	−	−	−	+	−
Pachytriton	+	−	−	−	−	−	+	+	−	−	+	+	+	+	+	+	−	−	−	−	−	−	−	−	+	−	+	+	−
AMBYSTOMATIDAE																													
Dicamptodon	+	−	−	−	−	−	+	−	−	−	+	+	−	−	+	−	−	−	+	+	+	−	−	−	−	+	−	−	−
Rhyacotriton	+	−	−	−	−	−	+	−	−	−	+	+	−	+	−	−	−	−	+	+	+	−	−	−	−	−	−	−	+
Rhyacosiredon	+	−	+	−	−	−	+	−	−	−	+	+	−	−	−	−	−	−	+	+	+	−	−	−	−	−	−	−	−
Ambystoma	+	−	+	−	−	−	+	−	−	−	+	+	+	+	+	+	+	+	+	+	+	+	−	+	−	+	−	−	−
HYNOBIIDAE																													
Hynobius	+	−	−	−	−	−	+	−	−	−	+	+	−	−	−	−	−	−	−	+	+	−	−	−	+	−	−	−	−
Batrachuperus	−	−	−	−	−	−	−	−	−	−	−	−	−	−	−	−	−	−	−	−	−	−	−	−	−	−	−	−	−
Ranodon	−	−	−	−	−	−	−	−	−	−	−	−	−	−	−	−	−	−	−	−	−	−	−	−	−	−	−	−	−
Onychodactylus	−	−	+	−	−	−	−	−	−	−	−	−	−	−	−	−	−	−	−	−	−	−	−	−	−	−	−	−	−

PLETHODONTIDAE	a	b	c	d	e	f	g	h	i	j	k	l	m	n	o	p	q	r	s	t	u	v	w	x	y	z	aa	bb	cc
Desmognathus	-	-	-	-	-	-	-	-	-	-	-	-	-	-	-	+	-	+	+	-	-	-	-	-	-	-	-	-	-
Phaeognathus	-	-	-	-	-	-	-	-	-	-	-	-	-	-	-	-	-	-	-	-	-	-	-	-	-	-	-	-	-
Gyrinophilus	+	-	-	+	+	+	-	-	+	+	+	-	-	-	-	-	-	-	-	-	-	-	-	-	-	-	-	-	-
Pseudotriton	+	-	-	+	+	+	-	-	+	+	+	-	-	-	-	-	-	-	-	-	-	-	-	-	-	-	-	-	-
Eurycea	+	-	+	+	+	+	-	+	+	+	-	-	-	-	-	-	-	-	-	-	-	-	-	-	-	-	-	-	-
Typhlotriton	+	-	+	-	+	+	-	+	+	+	-	-	-	-	-	-	-	-	-	-	-	-	-	-	-	-	-	-	-
Hemidactylium	+	-	+	-	+	+	-	+	+	+	-	-	-	-	-	-	-	-	-	-	-	-	-	-	-	-	-	-	-
Aneides	+	-	+	-	+	+	+	-	+	+	+	-	-	-	-	-	-	+	+	-	-	-	-	-	-	-	-	-	-
Plethodon	+	-	+	-	+	+	+	-	+	+	-	-	-	-	-	-	-	+	+	-	-	-	-	-	-	-	-	-	-
Ensatina	+	-	+	-	+	+	-	-	+	+	-	-	-	-	-	-	-	+	-	-	-	-	-	-	-	-	-	-	-
Batrachoseps	-	-	-	-	-	-	+	-	-	-	-	-	-	-	-	-	-	-	-	-	-	-	-	-	-	-	-	-	-
Hydromantes	+	-	-	+	+	+	-	-	+	-	+	-	-	-	-	-	-	+	-	-	-	-	-	-	+	-	-	-	-
Bolitoglossa	+	-	-	-	+	+	-	-	+	+	-	-	-	-	-	-	-	+	-	-	-	-	-	-	-	-	-	-	-
Chiropterotriton	+	-	-	-	-	-	-	-	+	-	-	-	-	-	-	-	-	+	-	-	-	-	-	-	-	-	-	-	-
Lineotriton	-	-	-	-	-	-	-	-	-	-	-	-	-	-	-	-	-	-	-	-	-	-	-	-	-	-	-	-	-
Oedipina	+	-	-	-	+	+	-	-	+	-	-	-	-	-	-	-	-	+	-	-	-	-	-	-	-	-	-	-	-
Parvimolge	-	+	-	-	-	-	-	-	-	-	-	-	-	-	-	-	-	-	-	-	-	-	-	-	-	-	-	-	-
Pseudoeurycea	+	-	+	-	+	+	-	-	+	+	-	-	-	-	-	-	-	+	-	-	-	-	-	-	-	-	-	-	-
Thorius	+	-	-	-	-	-	-	-	+	+	-	-	-	-	-	-	-	+	-	-	-	-	-	-	-	-	-	-	-

a. noxious skin secretion; b. toxic skin secretions; c. parotoid glands; d. glandular lateral warts; e. glandular tail dorsum; f. glandular tail venter; g. glandular body middorsum; h. secretion sprayed; i. aposematic coloration on dorsum; j. aposematic coloration on venter; k. pseudoaposematic coloration; l. immobile posture; m. body coiled; n. body flipped; o. venter exposed, tail up; p. venter exposed, chin up; q. rolls onto back; r. body arched; s. head butted; t. tail lashed; u. tail wagged; v. tail undulated; w. ribs pierce skin; x. hook on quadrate; y. vocalize; z. bite; aa. tail autotomy; bb. frontosquamosal arch; cc. expanded neural spine.

Glandular Tail Venter

This is found as a row of one or two large granular glands along the midventral line of the tail (Brodie et al., in press).

Glandular Body Middorsum

This refers to one row of enlarged granular glands along either side of the middorsal line of the body.

Secretion Sprayed

This refers to the condition where the secretion is forcibly ejected from the granular glands.

Aposematic Coloration on Dorsum

This is the presence of a potentially aposematic (warning) color pattern on the dorsal surface. Poulton (1890) suggested that the effective aposematic color combinations were white, yellow, red, or orange on black or a metallic coloration. These color patterns are thought to be most effective as alternating rings, longitudinal stripes, or spots (Poulton, 1890). The presence of one of the above color patterns in any member of a genus is indicated as positive in Table 1.

Aposematic Coloration on Venter

This refers to the presence of one of the aposematic color patterns listed under (i) on some part of the ventral surface of some member of the genus.

Pseudoaposematic Coloration

This refers to a color pattern that mimics that of a more noxious salamander. It is possible for a moderately noxious species to mimic a highly noxious one. Only mimetic relationships that have been demonstrated experimentally are included.

Immobile Posture

This occurs when a salamander becomes immobile (often but not necessarily in some distinctive posture). The body is held immobile during most antipredator displays.

Body Coiled

This posture is when the body is coiled with the snout positioned near or under the vent region. The body is held immobile but the tail may assume a number of positions, from being coiled over the head to held vertically and undulated.

Body Flipped

This refers to rapid protean movements that are associated with changes in shape, position and location of the salamander. This flipping is accomplished in different ways by different species. The most common type of flipping is the coil-uncoil flip in which the body is propelled by a series of rapidly alternating coiling and uncoiling movements. The limbs are clasped along the body and the body is held rigid during this type of flipping. A second common type of flipping is the running flip in which the tail is used to propel the body while the animal is running. This type of flip looks like the salamander is jumping erratically. The third and least common type of flipping is the serpentine flip in which the body and tail are held stiff and writhed laterally. This propels the salamander to the side and causes it to resemble a small snake. This latter type of flip has been observed only in the genus *Oedipina*.

Venter Exposed, Tail Up

This is when the tail is held in such a position as to expose the undersurface. The body usually is held rigid in either a coiled or uncoiled position.

Venter Exposed, Chin Up

This is when the head is held vertically so as to expose the chin. The body is held rigid in either a coiled or uncoiled position. This positioning of the body is apparently restricted to members of the Salamandridae and is a primary component of the behavior pattern widely referred to in the literature as the "unken reflex".

Rolls Onto Back

This is when a salamander rolls onto its back and becomes immobile in that position. This behavior pattern has been reported only for *Paramesotriton chinensis*.

Body Arched

This is when the midbody is elevated off the ground and is higher than either the pelvic or pectoral regions. The legs usually are extended lifting the entire body as high off the substrate as possible.

Head Butted

This is when the head is flexed vertically downward and the parotoid region is swung into the attacking predator.

Tail Lashed

This is when the tail is swung forcibly into the attacking predator. The tail is usually, but not always, held in an elevated position prior to being lashed.

Tail Wagged

This is when the tail is raised off the substrate, held straight, and swung from side to side. The tail is not swung forcibly nor necessarily toward a predator.

Tail Undulated

This is when the tail is moved in a sinuous manner. This is usually, but not always, exhibited with the tail held vertically.

Ribs Pierce Skin

This is when either the tip of the shaft of the rib or of an epipleural process penetrates through the lateral or dorsolateral skin of the body. The ribs may pierce through aglandular areas, as in *Pleurodeles waltl*, or through the glandular lateral warts, as in *Echinotriton andersoni*.

Hook on Quadrate

This is a sharp tipped, anteriorly curved spine projecting laterally from each quadrate. It lies within and pierces through a concentration of enlarged granular glands. This characteristic is found only in *Echinotriton*.

Vocalize

This refers to any sound emitted by salamanders in response to predator attack. Several different sounds have been described, including: squeaks, clicks, hisses, and barks.

Bite

This refers to a salamander biting a predator when approached or attacked.

Tail Autotomy

This refers to those salamanders that have anatomical specializations for tail loss (Wake and Dresner, 1967).

Frontosquamosal Arch

This refers to a complete bony arch over the eye from the frontal bone to the squamosal bone (Naylor, 1978).

Expanded Neural Spine

This refers to the series of interlocking plates that cap the neural spines of the vertebrate (Naylor, 1978).

CORRELATED ANTIPREDATOR MECHANISMS

Patterns of antipredator behaviors, and in many cases, coloration are correlated with the presence and distribution of granular glands (Brodie, 1977; Brodie et al., in press; Brodie and Gibson, 1969; Hubbard, 1903). The patterns of antipredator behaviors are also correlated with body form (e.g., stocky or slender) and specializations for tail autotomy. Certain sets of correlated antipredator mechanisms occur in several to many different species, and the term "suite" is used here to refer to those antipredator mechanisms that act together (probably synergistically) to protect a salamander from predator attack.

This is not a comprehensive listing of all correlated antipredator mechanisms found in terrestrial salamanders but is instead a presentation of only the most common suites of characters. There is some variability within each of these suites of characters, and individuals of a few species exhibit more than one of these suites of characters.

Suite I - Unken Reflex

This suite is characterized by a rigid immobile posture with the tail and chin elevated bright ventral coloration, and toxic skin glands that are generally evenly distributed over the dorsal surface. As far as is known, species exhibiting this suite of characters also have complete frontosquamosal arches and usually have expanded neural arches (Naylor, 1978).

This suite of characters is dependent on the presence of toxic skin secretions which render the salamander absolutely inedible to virtually all predators (Brodie et al., 1974a). This dependency on toxicity restricts the unken reflex to members of the Salamandridae where it has been described in the following genera: *Tylototriton, Echinotriton, Salamandrina, Cynops, Notophthalmus, Taricha, Triturus,* and *Paramesotriton*. All species examined from each of the above genera exhibited this suite of characters except *Triturus marmoratus* (Brodie, 1977; Brodie et al., in press).

Fig. 1. Low intensity unken reflex of *Echinotriton andersoni*, tail
undulated.

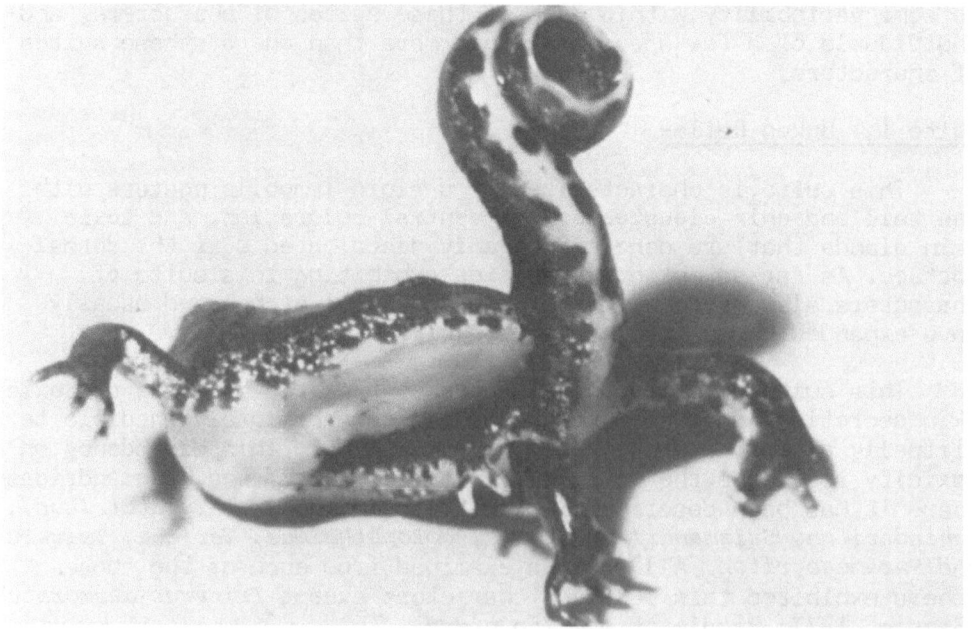

Fig. 2. High intensity unken reflex of *Triturus alpestris*.

There is some variability among the salamanders exhibiting Suite I antipredator mechanisms. Most species exhibit low-intensity antipredator behaviors in which the tail but not the chin is elevated (Fig. 1). Some species undulate the tail while in the low-intensity posture (e.g., *Echinotriton andersoni, Tylototriton verrucosus, Triturus alpestris,* and *Notophthalmus viridescens*). However, these do not undulate the tail during the high-intensity posture when the chin is elevated. Some species coil the body with the snout held near the vent region (Fig. 2) and other species keep the body straight (Fig. 3). The tail may be coiled or held straight.

All species exhibiting the unken reflex have the skin covering the dorsal surface abundantly supplied with granular glands. In addition some species have enlarged granular glands in localized regions of the skin (e.g., lateral warts, body middorsum, tail dorsum, and tail venter). These concentrations of enlarged granular glands may be associated with other antipredator behaviors.

One genus (*Echinotriton*) with this suite of antipredator mechanisms also has ribs and quadrate hooks that pierce the skin and are surrounded by enlarged granular glands (Nussbaum and Brodie, 1982; Brodie et al., in press). These sharp bones piercing the skin are not associated directly with Suite I characters.

It has been demonstrated that Suite I characters act synergistically to enhance the survival of these salamanders when attacked by avian predators (Johnson and Brodie, 1975). Birds learned to avoid *Taricha granulosa* offered with their bright orange ventral surface exposed more rapidly than the same salamanders with the dark dorsal surface showing. The rate of learning was even more rapid when birds were offered *T. granulosa* in the typical unken posture.

The toxicity and associated noxiousness protect salamanders with these characteristics from being eaten by most predators but do not prevent predators from attacking. The bright coloration and the distinctive unken posture which exhibits the ventral coloration are cues which are associated with toxicity and noxiousness by the predators (Johnson and Brodie, 1975). It has been shown that avian predators respond to color cues after contact with distasteful prey (e.g., Brodie and Brodie, 1980; Brodie and Howard, 1973; Howard and Brodie, 1971, 1973). The immobility aspect of the unken reflex reduces the intensity of predator attack (reviewed for salamanders, Brodie et al., 1974b). This reduction in attack intensity increases the probability that a predator will reject an inedible salamander without inflicting a serious wound. The presence of the bony frontosquamosal arch and expanded caps on the neural spines of the vertebrae also reduce the probability of injury from a predator that has not yet learned that a salamander is inedible (Naylor, 1978).

Fig. 3. High intensity unken reflex of *Taricha rivularis*.

Fig. 4. Tail lashing posture of *Tylototriton verrucosus*.

Suite II - Tail Lash

This suite is characterized by lashing of the tail, concentrations of enlarged granular glands along the dorsal surface of the tail, and well developed tail musculature. Aposematic coloration, when present, is on the dorsal surface.

Tail lashing is an active antipredator behavior pattern in which the tail is swung laterally with force toward any contact or an approaching predator. Tail lashing normally takes place from a posture with the hind legs extended, elevating the pelvic region high off the ground, and with the tail held in an arched position (Figs. 4, 5). The body is arched and held high off the ground in some species by the extension of both fore and hind legs (Fig. 6). In these species the tail is wagged slowly back and forth while the salamander postures, and is lashed from this position. Few species that lash the tail assume an antipredator posture with the body coiled.

This suite of characters is dependent on the presence of concentrations of enlarged granular glands along the dorsal surface of the tail. Most genera of terrestrial salamanders have at least some species with glandular concentrations along the tail dorsum (Table 1) and many but not all of these exhibit tail lashing. Many species with this morphology exhibit Suite III characteristics and a few exhibit Suite I characteristics instead of Suite II. Heavy tail musculature along with a concentration of glands on the tail dorsum is necessary for a salamander to effectively slap or lash the tail into a predator.

Members of each family of terrestrial salamanders exhibit Suite II characters, including the following: Salamandridae-*Tylototriton verrucosus*, *Pleurodeles waltl*, *Triturus marmoratus*; Hynobiidae-*Hynobius chinensis*; Ambystomatidae-*Dicamptodon ensatus*, *Rhyacotriton olympicus*, *Rhyacosiredon altamirani*, *Rhyacosiredon rivularis*, *Ambystoma gracile*, *A. lacustris*, *A. maculatum*, *A. mexicanum*, *A. opacum*, *A. ordinarium*, *A. talpoideum*, *A. tigrinum*; Plethodontidae-*Ensatina eschscholtzi*, *Plethodon glutinosus*, *P. jordoni*. It is likely that other members of the Hynobiidae also exhibit this suite of characters but that family has not been studied adequately.

The efficacy of Suite II characters in enhancing survival of predatory attacks has been demonstrated for *Ambystoma opacum* attacked by shrews, *Blarina brevicauda* (DiGiovanni and Brodie, 1981). Thirty *A. opacum* survived an average of 6.4 (range 2-12) one-minute trials during which they were attacked by shrews. Tail lashing was effective in repelling shrew attacks and the associated posture with the tail arched and wagged caused the shrews to attack the tail. The tail was attacked in 93.3% of the 194 trials compared to attacks to the body, 54.1%; legs 20.1%, and head, 13.9%. Directing an attacking predator to the tail is advantageous because the tail

Fig. 5. Tail lashing posture of *Ambystoma ordinarium*, note heavy granular secretion on tail and body.

Fig. 6. Tail lashing posture of *Ensatina eschscholtzi*, note secretions on tail.

contains the largest concentration of granular glands and is the most dispensable part of the salamander. As the volume of secretion was depleted after successive attacks the intensity of shrew attacks increased and the ability of *A. opacum* to repel the shrew decreased. This relationship confirmed the importance of Suite II characters acting together to enhance the ability of salamanders to repel predators.

Suite III - Tail Undulate

This suite is characterized by undulation of the tail, con-centrations of enlarged granular glands along the dorsal surface of the tail, and a long slender tail. Aposematic coloration, if present, is normally restricted to the dorsal surface, and most species have specializations for tail autotomy.

Undulating the tail is a passive antipredator behavior in which the tail is moved in a sinuous manner while the body is held immobile. The tail normally is held in a vertical position while undulated and the body usually is coiled with the head under the base of the tail. Species having this suite of characters may turn and lean the body and tail toward an attacking predator and may even swing the tail in that direction but they do not lash the tail with force.

This suite of characters (like Suite II) is dependent on the presence of a glandular tail ridge. An equally important structure for Suite III is the presence of a long slender tail. Many species exhibiting the unken reflex (Suite I) have a low-intensity posture very similar to Suite III; these species, however, change from tail undulation to a rigid unken with increased predator attack.

Some members of the Salamandridae, Ambystomatidae, and Pletho-dontidae exhibit Suite III characters. The only salamandrid known to exhibit this suite of characters is *Mertensiella caucasica* (Fig. 7), although as noted above many members of this family exhibit these behavior patterns in a low-intensity antipredator posture.

Seven members of the genus *Ambystoma* (Ambystomatidae) exhibit Suite III characters; these are: *A. annulatum, A. cingulatum, A. jeffersonianum, A. laterale, A. mabeei, A. macrodactylum,* and *A. texanum.* These species swing the tail into an attacking predator from the elevated and undulated posture (Fig. 8) but not with the force or accuracy of members of this genus listed under Suite II.

Suite III characters are most common among the Plethodontidae (Figs. 9, 10). All members of the following genera that have been investigated have this suite of characters: *Gyrinophilus, Pseudo-triton, Eurycea, Typhlotriton, Hemidactylium, Aneides, Hydromantes, Bolitoglossa, Chiropterotriton, Oedipina, Pseudoeurycea,* and *Thorius.* In addition western and small eastern members of the genus *Plethodon* have this suite of characters.

Fig. 7. Tail undulated posture of *Mertensiella caucasica*.

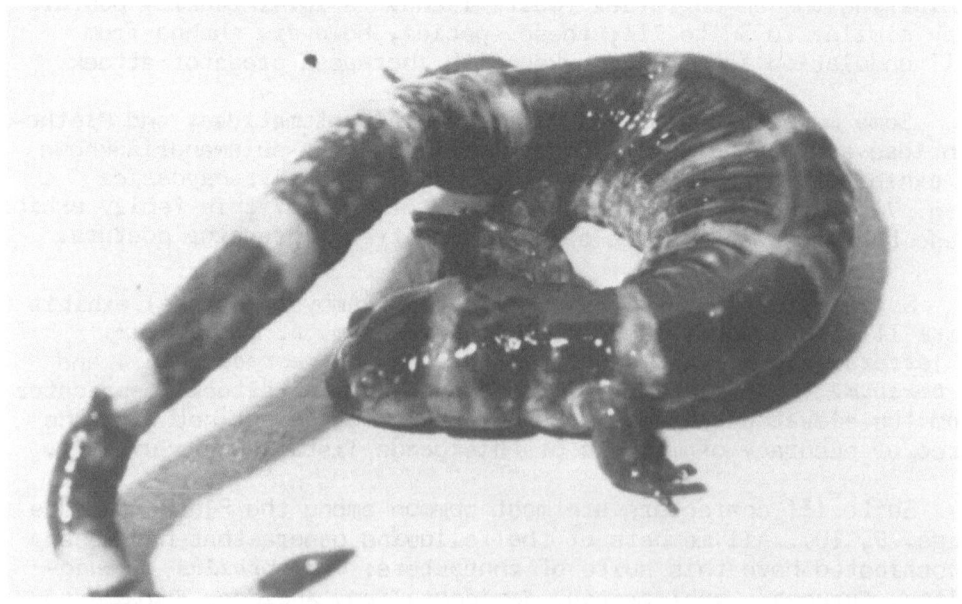

Fig. 8. Tail undulated posture of *Ambystoma annulatum*.

Fig. 9. Tail undulated posture of *Eurycea lucifuga*.

Fig. 10. Tail undulated posture of *Bolitoglossa occidentalis*.

Suite III characters act to protect salamanders by directing
the attacking predator to the tail which contains large amounts of
distasteful secretion and in most cases may be autotomized. Predators
are attracted to autotomized tails, which continue to undulate, and
during this time the salamander has a chance to escape (Wake and
Dresner, 1967; Ducey and Brodie, in press).

Suite IV - Head Butt

This suite is characterized by butting the head, concentrations
of enlarged granular glands in the parotoid region, and heavy bodies.
Aposematic coloration, if present is located on the dorsal surface.
The body is not coiled and the tail is not used in this suite of
characters.

Head butting is when the head is flexed downward and the back
of the head, bearing well developed parotoid glands, is swung or
lunged into a predator (Figs. 11, 12). The body is held high off
the substrate and the head and forebody are leaned toward the
predator. Most species vocalize while head butting. In several
species the parotoid glands are brightly colored with spots of
yellow or orange and the spotted pattern continues along the body
and tail.

Not all species with parotoid glands head butt. A number of
species flex the head down or hold the head flat to the substrate
when attacked but do not butt the head. These species are not considered to have Suite IV characters.

Suite IV characters are known for only a few species of the
Salamandridae and Ambystomatidae. The salamandrids *Tylototriton
verrucosus, Pleurodeles waltl, Salamandra salamandra,* and *S. atra*
exhibit this suite of characters, and it is probable that
Mertensiella luschani also has this suite of characters given its
glandular distribution. The ambystomatids *Ambystoma gracile,
A. maculatum,* and *A. talpoideum* also exhibit this suite of characters.
Many hynobiids reportedly have glandular concentrations in the
parotoid region and should be examined for this suite of characters.

Head butting the parotoid gland into the attacking predator
presents the predator with the most distasteful part of the
salamander and is effective in repulsing shrews (Brodie et al., 1979).

ORIGIN OF ANTIPREDATOR MECHANISMS

As presented in the previous section and Table 1 the most
common suites of correlated antipredator mechanisms occur in a
wide variety of terrestrial salamanders. Suites II, III, and IV
appear in more than one family, and, unless the common ancestor

Fig. 11. Head butted posture of *Ambystoma maculatum*, note secretion
on parotoid gland and body.

Fig. 12. Head butted posture of *Ambystoma talpoideum*.

of extant salamanders had all of the suites fully developed, this appearance in multiple families represents convergence between species.

Given the ubiquitous presence of granular glands among the living amphibia it is highly probable that ancestral salamanders had granular glands. Since the granular glands of living species (in cases studied) produce noxious secretions in frogs (e.g., Dodd, 1976; Formanowicz and Brodie, 1979, 1982; Phisalix, 1922) and caecilians (Brodie and Nussbaum, unpub. obser.) as well as salamanders, it is reasonable to predict that the granular glands of ancestral salamanders also produced slightly to moderately noxious secretion. Very little knowledge is available regarding the chemistry of glandular secretions of salamanders (e.g., Brandon and Huheey, 1981; Habermehl, 1971; Mosher et al., 1964) and it is not possible to determine if similarities exist between the secretions of different families.

It is possible to envision the evolution of each of the four suites of antipredator characters considered in this paper from a hypothetical ancestral salamander with granular glands (containing somewhat noxious secretions (diffusely distributed throughout the skin).Evolution of different specializations of the skin glands would provide the impetus for the evolution of each of the suites of characters (Fig. 13).

Suite I characteristics could have evolved from the hypothetical ancestor with the evolution of extreme repulsiveness (to the point of being toxic if eaten by predators) and granular glands abundantly distributed over the body. Low-intensity postures usually consist of placing the head under the tail and undulating the tail. These behavior patterns change to a rigid posture with the tail and chin up (exposing bright ventral coloration) after further predator attacks. These rigid immobile postures would be adaptive only in a salamander that was inedible to predators (Brodie et al., 1974b).

Most salamanders with Suite I characters have slight concentrations of granular glands along the dorsal surface of the tail and in the parotoid region (Table 1). These concentrations usually are not used in any discernable manner during predator attacks and may have evolved prior to the development of the extreme noxiousness now found in salamanders with this suite of characters. These species are the ones that exhibit undulation of the tail in a low-intensity display. These species are also less toxic than those salamanders that lack glandular concentrations and do not undulate their tails (Brodie, 1982; Brodie et al., 1974a).

Only one species with Suite I characters also exhibits other suites. This is *Tylototriton verrucosus* which also exhibits Suites II and IV. The presence of this diversity in one species could be used to argue that ancestral salamandrids had several suites well developed and certain suites were lost in some linages. It is un-

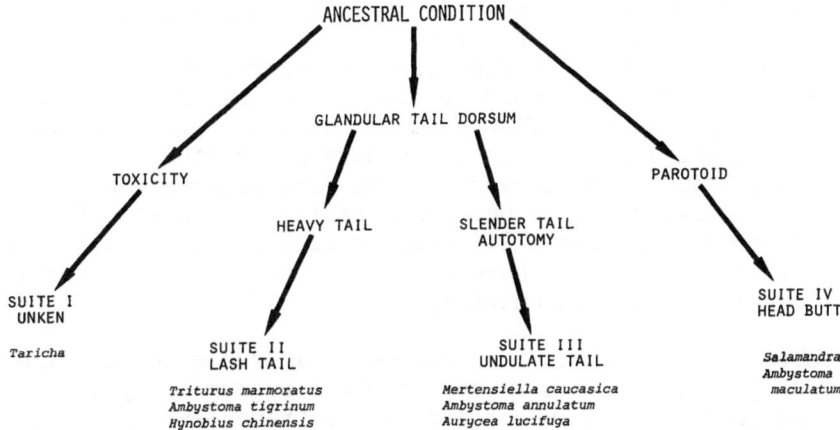

Fig. 13. Probable evolutionary divergence of antipredator mechanisms
    from the hypothetical ancestral condition to living
    salamanders. One representative species of each family
    exhibiting a suite of characters is indicated.

likely that this point of view is correct because Suite I characters
are not compatible with those of the other suites and the single
example (*T. verrucosus*) does not have any suite of characters fully
expressed.

    Certain concentrations of granular glands would be advantageous
even without the behavior patterns associated with them in living
salamanders. Predators most often attack the head-end of prey (Curio,
1976) and salamanders (even fully palatable species) twist the tail
over the head when seized on the head, neck, or body by a predator
(Brodie, unpub. obser.). It is reasonable to predict that ancestral
salamanders had similar behavior patterns. These behavior patterns
by salamanders and their predators would cause a selective advantage
to any individual with a concentration of noxious secretions along
the tail, or in the parotoid region even without the associated
specialized patterns of behavior of Suites II, III, IV. The presence
of concentrations of granular glands would cause salamanders to be
preadapted for any behavior patterns that would increase the efficacy
of these glands in repulsing predators.

    Concentrations of granular glands along the dorsal surface of
the tail are necessary for the presence of either Suite II or Suite
III characters. As different taxa entered different adaptive zones
and were subjected to different environmental pressures they also
specialized further in their antipredator mechanisms. The initial
evolution of heavily muscled or slender tails probably had more to
do with the physical environment than with predator pressure but
the shape of the tail dictated how the tail could be used as an
antipredator device (Fig.

Suite II characteristics are probably most similar to those of the ancestral condition. As the contents of the granular glands became more distasteful or the volume of secretion along the dorsal surface of the tail increased (or both) the tail became a more effective weapon in repulsing predator attacks. Some species exhibit tail lashing while walking (notably several species of *Hynobius* and *Ambystoma*). This is not very different from wagging the tail or swinging the tail over the head when seized as outlined for the ancestral condition. Suite III characteristics have evolved primarily in those groups that have developed long slender tails and specializations for tail autotomy.

Suite IV characteristics involve an increase in the size and number of granular glands in the parotoid region as compared to the ancestral condition. Only species with extreme development of the parotoid glands exhibit head butting behavior. Several species exhibiting Suite IV characteristics when attacked on the head or front legs exhibit Suite II characteristics when attacked on the tail or rear legs.

The presence of similar suites of characters in nonsalamanders would support the contention that these suites are adaptive and antipredator in function. A number of frogs or lizards exhibit each of the four suites of antipredator characters considered in this paper, The following are not exhaustive lists but indicate the generality of the suites of characters described for salamanders.

Suite I characters are found in members of the genus *Bombina* (Hinsche, 1926) and *Melanophryniscus stelzneri* (Cochran, 1961; McDiarmid, 1971). These frogs are toxic, have bright ventral coloration, and expose the ventral coloration in a rigid unken reflex when attacked by predators.

A number of spiny tailed lizards (e.g., *Uromastix* and *Cordylus*) elevate the body and lash the tail when approached or attacked by predators (Bellairs, 1970). This behavior pattern is the same as that of Suite II but the repellent is mechanical, in the form of spines, rather than distasteful secretions.

Many lizards elevate and undulate the tail when contacted by a predator (Bellairs, 1970; Johnson and Brodie, 1974); these lizards generally do not have distasteful secretions but do exhibit tail autotomy. At least some geckos of the genus *Diplodactylus* do have distasteful secretions concentrated in the tail (Bustard, 1964) giving them the characters of Suite III.

Toads of the genus *Bufo* are the most obvious animals exhibiting Suite IV characters but the head spines of some lizards are used in a head butt similar to that of salamanders and even some snakes have distasteful secretions concentrated in the neck region (Bellairs, 1970)

If the above scenario about the evolution of the four suites of antipredator characteristics being considered is correct, the families of terrestrial salamanders have evolved parallel specializations to protect themselves from predators. Members of the genera and species of these families have then undergone remarkable convergences of antipredator mechanisms.

ACKNOWLEDGEMENTS

I thank J. A. Johnson, R. Lund, J. J. Napolitano and R. A. Nussbaum who reviewed earlier versions of this paper. My research on amphibian antipredator mechanisms has been made possible by support from the following: American Philosophical Society, E. N. Huyck Preserve, Highlands Biological Station, National Geographic Society, National Science Foundation, and Sigma Xi.

REFERENCES

Arnold, S. J., 1982, A quantitative approach to antipredator performance: salamander defence against snake attack, Copeia, 1982:247-253.

Bellairs, A., 1970, "The life of reptiles," Vol. II, Universe Books, New York.

Brandon, R. A., and Huheey, J. E., 1981, Toxicity in the plethodontid salamanders Pseudotriton ruber and Pseudotriton montanus (Amphibia, Caudata), Toxicon, 19:25-31.

Brandon, R. A., Labanick, G. M., and Huheey, J. E., 1979a, Relative palatability, defensive behavior, and mimetic relationships of red salamanders (Pseudotriton ruber), mud salamanders (Pseudotriton montanus), and red efts (Notophthalmus viridescens), Herpetologica, 35:289-302.

Brandon, R. A., Labanick, G. M., and Huheey, J. E., 1979b, Learned avoidance of brown efts, Notophthalmus viridescens louisianensis (Amphibia, Urodela, Salamandridae), by chickens, J. Herpetol., 13:171-176.

Brodie, E. D. Jr., 1977, Salamander antipredator postures, Copeia, 1977:523-535.

Brodie, E. D. Jr., 1978, Biting and vocalization as antipredator mechanisms in terrestrial salamanders, Copeia, 1978:127-129.

Brodie, E. D. Jr., 1982, Toxic salamanders, JAMA, 247:1408.

Brodie, E. D. Jr., Antipredator adaptation of neotropical salamanders (supergenus Bolitoglossa, family Plethodontidae), Nat. Geo. Soc. Res. Reports, 14 , in press.

Brodie, E. D. Jr., and Brodie, E. D. III., 1980, Differential avoidance of mimetic salamanders by free-ranging birds, Science, 208:181-183.

Brodie, E. D. Jr., and Formanowicz, D. R. Jr., 1981, Larvae of the predaceous diving beetle Dytiscus verticalis acquire an avoidance response to skin secretions of the newt *Notophthalmus viridescens*, Herpetologica, 37:172-176.

Brodie, E. D. Jr., and Gibson, L. S., 1969, Defensive behavior and skin glands of the northwestern salamander, *Ambystoma gracile*, Herpetologica, 25:187-194.

Brodie, E. D. Jr., Hensel, J. L., and Johnson, J. A., 1974a, Toxicity of the urodele amphibians *Taricha, Notophthalmus, Cynops,* and *Paramesotriton* (Salamandridae), Copeia, 1974:506-511.

Brodie, E. D. Jr., and Howard, R. R., 1973, Experimental study on Batesian mimicry in the salamanders *Plethodon jordani,* and *Desmognathus ochrophaeus*, Amer. Midl. Nat. 90:38-46.

Brodie, E. D. Jr., Johnson, J. A., and Dodd, C. K. Jr., 1974b, Immobility as a defensive behavior in salamanders, Herpetologica, 30:79-85.

Brodie, E. D. Jr., Nowak, R. T., and Harvey, W.R., 1979, The effectiveness of antipredator secretions and behavior of selected salamanders against shrews, Copeia, 1979:270-274.

Brodie, E. D. Jr., Nussbaum, R. A., and DiGiovanni, M., Antipredator adaptations of Asian salamanders (Salamandridae) compared to other members of the family, Herpetologica, in press.

Bustard, H. R., 1964, Defensive behavior shown by Australian geckos, genus *Diplodactylus*, Herpetologica, 20:198-200.

Cochran, D. M., 1961,"Living amphibians of the world," Hamish Hamilton Ltd., London.

Curio, E., 1976, "The ethology of predation," Springer-Verlag, Berlin.

DiGiovanni, M., and Brodie, E. D. Jr., 1981, Efficacy of skin glands in protecting the salamander *Ambystoma opacum* from repeated attacks by the shrew *Blarina brevicauda*, Herpetologica, 37: 234-237.

Dodd, C. K. Jr., 1976, A bibliography of anuran defensive mechanisms, Smithsonian Herp. Info. Ser., 37:1-10.

Dodd, C. K. Jr., and Brodie, E. D. Jr., 1976, Defensive mechanisms of neotropical salamanders with an experimental analysis of immobility and the effect of temperature on immobility, Herpetologica, 32:269-290.

Dodd, C. K. Jr., Johnson, J. A., and Brodie, E. D. Jr., 1974, Noxious skin secretions of an eastern small *Plethodon, P. nettingi hubrichti*, J. Herpetol ., 8:89-92.

Ducey, P. K., and Brodie, E. D. Jr., Salamanders respond selectively to contacts with snakes: survival advantage of alternative antipredator strategies, Copeia, in press.

Eber, G., 1954, Über die Morphologie von Hautdrüsenansammlungen bei Salamandriden, Zool. Anz., 152: 153-162.

Formanowicz, D. R. Jr., and Brodie, E. D. Jr., 1979, Palatability and antipredator behavior of selected *Rana* to the shrew *Blarina*, Am. Midl. Nat., 101:456-458.

Formanowicz, D. R. Jr., and Brodie, E. D. Jr., 1982, Relative palatabilities of members of a larval amphibian community, Copeia, 1982:91-97.

Habermehl, G., 1971, Toxicology, pharmacology, chemistry, and biochemistry of salamander venom, in:"Venomous animals and their venoms," W. Bücherl and E. Buckley, eds., Academic Press, New York.

Hensel, J. L. Jr., and Brodie, E. D. Jr., 1976, An experimental study of aposematic coloration in the salamander *Plethodon jordani*, Copeia, 1976:59-65.

Hinsche, G., 1926, Vergleichende untersuchungen zum sogenannten unkenreflex, Biol. Zentralbl., 46:296-305.

Howard, R. R., and Brodie, E. D. Jr., 1971, Experimental study of mimicry in salamanders involving *Notophthalmus viridescens viridescens* and *Pseudotriton ruber schencki*, Nature, 233:277.

Howard, R. R., and Brodie, E. D. Jr., 1973, A Batesian mimetic complex in salamanders: responses of avian predators, Herpetologica, 29:33-41.

Hubbard, M. E., 1903, Correlated protective devices in some California salamanders, Univ. California Pub. Zool., 1:157-168.

Johnson, J. A., and Brodie, E. D. Jr., 1974, Defensive behaviour of the western banded gecko, *Coleonyx variegatus*, Anim. Beh., 22:684-687.

Johnson, J. A., and Brodie, E. D. Jr., 1975, The selective advantage of the defensive posture of the newt, *Taricha granulosa*, Am. Midl. Nat., 93:139-148.

McDiarmid, R. W., 1971, Comparative morphology of frogs of the neotropical genera *Atelopus, Dendrophryniscus, Melanophryniscus,* and *Oreophrynella*, Bull. L.A. Co. Mus. Nat. Hist., 12:1-66.

Mosher, H. S., Fuhrman, F. A., Buckwald, H. D., and Fisher, H. G., 1964, Tarichatoxin-tetrodotoxin: a potent neurotoxin, Science, 144:1100-1110.

Naylor, B. G., 1978, The frontosquamosal arch in newts as a defence against predators, Can. J. Zool., 56:2211-2216.

Nowak, R. T., and Brodie, E. D. Jr., 1978, Rib penetration and associated antipredator adaptations in the salamander *Pleurodeles waltl* (Salamandridae), Copeia, 1978:424-429.

Nussbaum, R. A., and Brodie, E. D. Jr., 1982, Partitioning of the salamandrid genus *Tylototriton* Anderson (Amphibia:Caudata) with a description of a new genus, Herpetologica, 38:320-332.

Phisalix, M., 1922, "Animaux venimeux et venins, "Vol II, Masson, Paris.

Poulton, E. B., 1890, "The colours of animals," Kegan Paul, Trench, Trübner, and Co. Ltd., London.

Wake, D. B., and Dresner, I. G., 1967, Functional morphology and evolution of tail autotomy in salamanders, J. Morph., 122:265-305.

THE INTER-NESTING INTERVAL OF ZAKYNTHOS LOGGERHEADS

D. Margaritoulis

Ministry of Physical Planning, Housing and Environment

Athens, Greece

INTRODUCTION

Although sea turtles live almost all their lives in the sea, they depend on the terrestrial habitat for reproduction. Female turtles come on land for short periods to lay their eggs and hatchlings emerging from nests in the sand walk briefly on land before entering (probably forever if they are males) the marine habitat. Nesting occurs on specific beaches, the "nesting beaches" or "rookeries", where the majority of the female population converges during the nesting season.

The loggerhead sea turtle *Caretta caretta* (L.) nests in various beaches along the southern and western coasts of Greece, but the most important rookery seems to be on the island of Zakynthos. Nesting in Zakynthos concentrates on Sekania, a small beach on the southern coast (Margaritoulis, 1982). This is perhaps the last active rookery on the island, for the other beaches suffer high pressures from development and tourism. During the nesting season (June until August) female loggerheads come ashore at night and, following a stereo-typed behavior, excavate a nest in the sand where they deposit about 100 eggs.

Despite the fact that emerging females leave conspicuous tracks on beaches, it is not feasible to estimate exactly the size of the nesting population by counting tracks for two reasons. First, sea turtles quite often come ashore without laying, and second, they may nest more than once in the same season.

Non-nesting emergences vary from a simple walk on the beach with no excavation to a series of nearly completed nests. The reason

for a non-nesting emergence is not always clear. Sometimes various
obstructions may hinder nesting,but often the turtle returns to the
water for no apparent cause. However, it is not always easy to
distinguish a non-nesting emergence from a nesting one by inspecting
the spoor. This is mainly caused by the number of superficial ex-
cavations a turtle may make in the course of an emergence, regard-
less of whether she is going to lay or not. As a matter of fact,
even a seemingly straightforward nesting spoor does not provide one
hundred per cent evidence that the turtle has laid.

It seems that fishermen and other people associated with sea
turtles knew long ago that loggerheads, as all other species of
sea turtles, may nest more than once in the same season (Mast, 1911;
Carr, 1952). Multiple nesting in loggerheads, first verified by
Caldwell et al. (1959) in Georgia, has been widely documented in
most loggerhead rookeries around the world. As far as can be
ascertained, in the Mediterranean region such work has been presented only for the southern coast of Turkey (Geldiay, 1980).

In Zakynthos the first evidence of multiple nesting was groups
of unlaid eggs of several distinct sizes seen in the dissection of
a turtle found dead on a nesting beach in 1978 (Margaritoulis, in
press).

A long-term tagging program was initiated in the summer of
1982 to assess various parameters of Zakynthos loggerhead population.
This paper presents mainly a preliminary result of the tagging
effort: an estimation of the inter-nesting interval; e.g., the
elapsed time between successive nestings of an individual during
the same season.

METHODS

Tagging of the turtles started on 25 June and continued until
the end of August when the number of emergences had decreased
considerably. During this period patrolling was done on parts of
the Western Peloponnesus coast in order to detect whether a turtle,
tagged in Zakynthos, would appear on the Peloponnesus. Tagging and
patrolling was done mostly by students working singly or in teams.

The tags used were monel self-piercing tags purchased from
the National Band and Tag Company by the Hellenic Society for the
Protection of Nature. Each tag bears a serial number and the
Society's address to facilitate return.

Great care was taken to avoid disturbing the nesting females.
For this reason tags were usually applied during or after ovi-
position when turtles exhibit the least sensitivity to disturbance.
The tags were applied to the posterior edge of the left front

flipper and subsequently were carefully inspected to check whether the locking mechanism had properly clipped.

A nesting emergence was scored as successful only if eggs were actually seen. Turtles encountered during nest covering or during their return to the sea were recorded as "no data", although they might have nested.

To avoid confusion, dates during the study period were not changed after 24.00 h but were kept the same until dawn (06.00 h).

RESULTS AND DISCUSSION

One hundred forty-eight females were tagged on Zakynthos, 97% of them on Sekania beach. Twenty-five individuals from the group tagged on Sekania were seen again from 1 to 6 times (38 recaptures) on the same beach (Table 1). Eleven additional females were tagged during patrols of the Western Peloponnesus coast, but these animals, were never seen again. No turtles tagged on Zakynthos were encountered on the Western Peloponnesus coast.

Only successful nestings are taken into account in calculating the inter-nesting interval. Figure 1 shows the distribution of intervals preceding each of the fourteen recorded successful returns. The prominent group, including intervals from 13 to 20 days, has an arithmetic mean of 14.6 days and apparently represents the turtles which nested successfully for a second (observed) time. Inclusion in this group of turtle Z-719 which renested after 20 days is justified by its history (Table 1).

Turtles Z-274 and Z-859 were observed nesting three successive times at intervals of 14-15 days and 13-14 days respectively (Table 1). The two individuals which nested after 29 and 33 days of their previous observed nesting (Fig. 1) may also represent 3-time nesters, their intermediate nesting having been missed by the observers. Behavior of the individual which renested after an interval of 3 days (Fig. 1) will be explained below.

Apart from helping estimate the inter-nesting interval, tagging can provide valuable information on other aspects of turtle behavior. Although the observers were instructed to avoid approaching turtles during the initial stages of nesting (when the animals are most sensitive), it seems that some females were disturbed and returned to the sea without laying. Unfortunately, no record was kept of turtles that were clearly disturbed by people, and therefore, non-nesting emergences resulting from human disturbance cannot be distinguished from the others,sea turtles often make. In any case, from Table 1 it is seen that turtles which have returned to the sea without laying, usually come back to nest either the same night

Table 1. In-season recovery data of loggerheads tagged in Zakynthos

	First tagged			Recaptures	
Tag No	Date tagged	Nest	Date	Days*	Nest
Z-251[a]	10.7.82	Yes	12.8.82	33	Yes
Z-255	10.7.82	Yes	10.8.82	31	No
Z-256	10.7.82	No	26.7.82	16	Yes
Z-261	11.7.82	Yes	30.7.82	19	?
Z-262	11.7.82	Yes	14.7.82	3	Yes
Z-267	11.7.82	Yes	28.7.82	17	?
Z-268	12.7.82	?	28.7.82	16	Yes
Z-274	12.7.82	Yes	26.7.82	14	Yes
			10.8.82	15	No
			10.8.82	0	Yes
Z-275	12.7.82	?	16.7.82	4	?
Z-279	13.7.82	Yes	11.8.82	29	No
			11.8.82	0	Yes
Z-282[b]	14.7.82	Yes	28.7.82	14	Yes
Z-286	16.7.82	?	23.8.82	38	Yes
Z-293	18.7.82	?	4.8.82	17	Yes
Z-300	22.7.82	Yes	5.8.82	14	Yes
Z-715	25.7.82	Yes	10.8.82	16	Yes
Z-719	25.7.82	Yes	11.8.82	17	No
			12.8.82	1	No
			12.8.82	0	No
			14.8.82	2	No
			14.8.82	0	No
			14.8.82	0	Yes
Z-790	3.8.82	Yes	5.8.82	2	?
			21.8.82	16	No
Z-814	5.8.82	No	5.8.82	0	No
			6.8.82	1	No
			6.8.82	0	Yes
Z-821	7.8.82	No	8.8.82	1	No
			8.8.82	0	No
Z-833	9.8.82	No	9.8.82	0	Yes
Z-851	12.7.82	Yes	26.7.82	14	Yes
Z-857	12.7.82	Yes	26.7.82	14	Yes
Z-859	13.7.82	Yes	26.7.82	13	Yes
			9.8.82	14	Yes
Z-869	15.7.82	Yes	28.7.82	13	Yes
Z-870	17.7.82	?	2.8.82	16	Yes

(*)days elapsed between emergences, (?)no data, (a)retagged with
Z-838, (b)double tagged with Z-283 on the right front flipper.

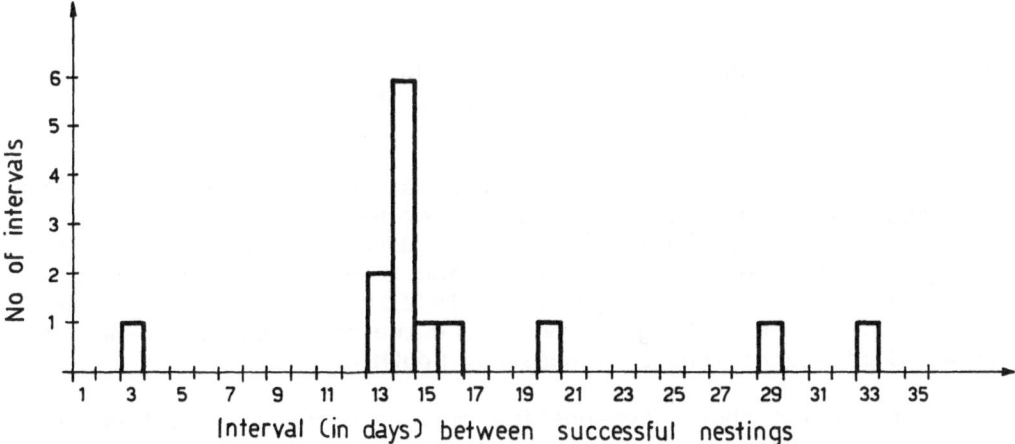

Fig. 1. Distribution of intervals preceding a successful nesting.

(3 cases) or on following nights (2 cases). The longest recorded interval between a non-nesting emergence and a nesting one was 3 days for turtle Z-719. This individual, tagged after egg-laying on 25 July, was observed in non-nesting emergences once on 11 August, twice on 12 August, and twice on 14 August, to nest again only after a third emergence on 14 August (Table 1). Turtle Z-275 might have exceeded this interval, emerging again after 4 days; but in both emergences there are no data to determine whether the individual nested or not. Hughes et al. (1967) have repeatedly disturbed an individual loggerhead which finally nested 11 days after the first observed non-nesting emergence.

Tag application greatly disturbed a number of turtles (N≅15), especially when done during oviposition. The usual reaction of the disturbed animals was an immediate interruption of egg-laying; oviposition usually resumed after a while. Some individuals, however, stopped laying and immediately covered the nest (although only superficially) and returned to the sea, presumably without having laid the full clutch. One of these turtles (Z-262) emerged again 3 days after the interrupted nesting and laid a small number of eggs, apparently the remainder of her clutch. Turtle Z-790 was tagged during oviposition and was disturbed to the point that she returned to the water so hastily the observers had to cover her nest. Her emergence 2 days later (although with an undetermined result) represents a probable continuation of the previously interrupted nesting. For this reason the recorded 3-day interval (Fig. 1) of the individual Z-262 was not included in the calculation. In the literature there are examples of loggerheads in Florida (2 cases) nesting twice in 2 days (Davis and Whiting, 1977; Worth and Smith, 1976) and in Georgia (1 case) in a non-nesting emergence 6 days

after a previous nesting (Caldwell et al., 1959). Although these
authors do not state whether the animals were disturbed during the
previously recorded egg-laying, it is thought that a second
emergence after such a short interval might be a result of disturbance.

The calculated inter-nesting interval for Zakynthos loggerheads
is comparable with those for other loggerhead rookeries with the
exception of Turkey (Table 2). Existing data from Zakynthos are not
enough to determine if the inter-nesting interval changes during
the laying season. Hughes and Brent (1972) have suggested that sea
temperature may affect egg ripening and therefore the inter-nesting
interval of loggerheads in Tongaland, South Africa.

Considering that a non-nesting emergence is an indication of
an imminent nesting, the observed total number of nests per female
is as much as these: two individuals (Z-274 and Z-859) were recorded
nesting three successive times and it is possible that four others
(Z-251, Z-255, Z-286 and Z-279) were also 3-time nesters (4% of the
tagged animals), fourteen turtles (9.5%) deposited probably two
clutches each, and the remaining 128 individuals (86.5%) laid
probably one clutch each.

Although the laying season in Zakynthos lasts for about 90 days
which could accommodate as many as 6 nestings of one female it is
improbable that this number of nests was made by any individual.
Nesting activity in Zakynthos increases during June, reaches a
maximum in the first two weeks of July, and then falls off gradually
to end in August (Margaritoulis, unpublished data). This nesting
pattern and the above findings suggests that most turtles do not
nest throughout the entire season but for only a part of it. Further
evidence is provided by the declining probability of recaptures as
the nesting season progresses. Intense tagging on Zakynthos began
on July 10th; e.g., during the second half of the nesting season.
It is seen from Table 3 that the majority of recaptures (37%) are
individuals tagged at Sekania during the interval 10-14 July which
coincides with the peak in nesting activity. Thereafter, the percentage of recaptures decreases rapidly, presumably as the turtles
depart from the breeding area.

The overall recapture rate of the present study (≈17%) compared
to similar investigations in other localities (Hughes and Brent, 1972;
Talbert et al., 1980) is considered low. This cannot be attributed
to relocation of in-season nesting on Zakynthos since the other
available beaches on the island are subject to heavy disturbances
which discourage most nesting activity. Relocation of nesting in
areas outside Zakynthos seems also unlikely according to the evidence
provided by the patrols on the Western Peloponnesus coast. Assuming
that Zakynthos loggerheads nest as many times per season as logger-

Table 2. Inter-nesting intervals of loggerhead populations in Zakynthos and in other localities

Locality	Source	Inter-nesting interval (in days)	
		Mean	Range
Zakynthos, Greece	present study	14.6	13-20
Köycegiz, Turkey	Geldiay, 1980	23.4	18-28
South Carolina, USA	Talbert et al., 1980	13.0	
Georgia, USA	Caldwell et al., 1959		12-15
Hutchinson, Florida, USA	Worth and Smith, 1976	14.0	11-17
Everglades, Florida, USA	Davis and Whiting, 1977	12.0	
Buritaca, Colombia	Kaufmann, 1975	15.0	13-20
Tongaland, S. Africa	Hughes and Brent, 1972	15.2*	
Queensland, Australia	Bustard, 1972	15.0	12-17

*Arithmetic mean calculated from values given by the authors for the seasons 1969/70-1970/71-1971/72.

Table 3. Numbers of turtles tagged during 5-day intervals and numbers
(and percentages) recaptured subsequently during the 1982
nesting season at Sekania, Zakynthos

Dates	Turtles tagged	Turtles subsequently recaptured during the season (%)
10.7-14.7	38	14 (37%)
15.7-19.7	19	4 (21%)
20.7-24.7	9	1 (11%)
25.7-29.7	22	2 (9.1%)
30.7- 3.8	10	1 (10%)
4.8- 8.8	27	2 (7.4%)
9.8-13.8	13	1 (7.6%)
14.8-18.8	5	0
19.8-23.8	1	0
24.8-28.8	0	0
Total	144	25 (17%)

heads in other localities, the observed low recapture rate is
probably an articraft due to the late beginning of tagging.

Multiple nesting represents a tremendous metabolic and physiological effort to the turtle (Urlich and Parkes, 1978). Assuming
that a loggerhead makes 4 nests of 100 eggs each and each egg weighs
40 g, over a period of 60 days a medium-sized loggerhead could
deposit the equivalent of 1/6 of her body weight in the form of
eggs. The survival value of such an effort is evident considering
that loggerheads, as other sea turtles, may migrate thousands of
miles between their rookery and feeding grounds (Bustard and Limpus,
1970; Bustard and Limpus, 1971; Hughes, 1974). The risk involved
in such long travels is probably manifested by the irregular re-
migration patterns and by the fact that a great percentage (40-50%)
of loggerheads may nest only during one season in their lifetime
(Hughes, 1976).

ACKNOWLEDGEMENTS

Financial assistance for the program was provided by the Greek
Ministry of Physical Planning, Housing and the Environment. The
Hellenic Society for the Protection of Nature bought the tagging
equipment.

Especial thanks are due to Dr. Jack Frazier of the Smithsonian
Institution who provided his expert "know-how" in solving various
problems and who with S. Salas tagged many turtles. Dr. J. Frazier

made also many valuable comments on the manuscript.

I am grateful to Prof. N.S. Margaris of the Laboratory of Ecology, University of Thessaloniki for his continued encouragement.

My personal thanks to the dedicated students who did most of
the tagging and patrolling of the beaches: D. Botsea, M. Gaethlich,
K. Kalogirou, P. Kanellopoulos, E. Karanatsidis, A. Kotitsas,
L. Loukas, K. Moraitis, M. Papoutsa, E. Roidou and A. Tselepidis.

The work at Zakynthos received the friendly and helpful interest
of the Prefecture of Zakynthos and of many local citizens to whom
I am grateful.

Finally, I am indebted to Helen O'Keefe for her editorial
assistance.

REFERENCES

Bustard, R., ed., 1972, "Sea turtles, natural history and conser-
    vation," Collins, London.
Bustard, R., and Limpus, C., 1970, First international recapture of
    an Australian tagged loggerhead turtle, Herpetologica, 26(3):
    358-359.
Bustard, R., and Limpus, C., 1971, Loggerhead turtle movements,
    Br. J. Herpet., 4(9):228-230.
Caldwell, D. K., Berry, F. H., Carr, A., and Ragotzkie, R. A., 1959,
    Multiple and group nesting by the Atlantic loggerhead turtle,
    Bull. Fla. St. Mus., 4(10):309-318.
Carr, A., ed., 1952, "Handbook of turtles," Cornell Univ. Press,
    Ithaca, New York.
Davis, G. E., and Whiting, M. C., 1977, Loggerhead sea turtle nest-
    ing in Everglades National Park, Florida, USA, Herpetologica,
    33(1):18-28.
Geldiay, R., 1980, Observation of the population dynamics and tagging
    procedures on the sea turtles (Caretta c. caretta L. and
    Chelonia m. mydas L.) of the Aegean and Mediterranean coasts
    of Turkey, Unpublished report to IUCN/WWF on project No 1419.
Hughes, G. R., 1974, The sea turtles of Southeast Africa. II. The
    biology of Tongaland loggerhead turtle Caretta caretta L.,with
    comments on the leatherback turtle Dermochelys coriacea L.
    and the green turtle Chelonia mydas L. in the study region.
    Oceanographic Research Institute Invest. Rep., No 36.
Hughes, G. R., 1976, Irregular reproductive cycles in the Tongaland
    loggerhead sea-turtle, Caretta caretta L., Zoologica Africana,
    11:285-291.
Hughes, G. R., Bass, A. J., and Mentis, M. T., 1967, Further
    studies on marine turtles in Tongaland, I., Lammergeyer, 7:
    1-55.

Hughes, G. R., and Brent, B., 1972, The marine turtles of Tongaland,
    VII., Lammergeyer, 17:40-62.
Kaufmann, R., 1975, Studies on the loggerhead sea turtle *Caretta
    caretta caretta* (Linné) in Colombia, South America, Herpeto-
    logica, 31:323-326.
Margaritoulis, D., 1982, Observations on loggerhead sea turtle
    Caretta caretta activity during three nesting seasons (1977-
    1979) in Zakynthos, Greece, Biol. Conservation, 24(3):193-204.
Margaritoulis, D., Observations on the breeding behavior and eco-
    logy of *Caretta caretta* L. in Zakynthos, Greece, in:"Proceed-
    ings of the 2nd International Congress on Zoogeography and
    Ecology pertaining to Greece and adjacent regions," in press.
Mast, S. O., 1911, Behavior of the loggerhead turtle in depositing
    its eggs. Pap. Tortugas Lab., Carnegie Inst. Washington, 3:
    63-67.
Talbert, O. R., Stancyk, S. E., Dean, J. M., and Will, J. M., 1980,
    Nesting Activity of the Loggerhead Turtle (*Caretta caretta*)
    in South Carolina. I.: A Rookery in Transition, Copeia, 1980(4):
    709-718.
Worth, D. W., and Smith, J. B., 1976, Marine turtle nesting on
    Hutchinson Island, Florida, in 1973, Fla. Dept. Nat. Res. and
    Mar. Res. Pub., 18.
Ulrich, G. F., and Parkes, A. S., 1978, The Green sea turtle
    (*Chelonia mydas*): further observations on breeding in cap-
    tivity, J. Zool., 185:237-251.

REPRODUCTION STRATEGIES IN BIRDS OF THE TROPICS

A. Chandola, M. Saklani, M. Bisht and D. Bhatt

Department of Zoology
Garhwal University
Srinagar (Garhwal) 246174, U.P., India

INTRODUCTION

For the successful survival of a species it is imperative that the animals be in perfect harmony with their environment. The environment is essentially cyclic in nature, the major periodicities deriving from the rotation of earth around its axis (daily changes) and around the sun (annual changes) and also from the rotation of moon around the earth (monthly changes). The adaptation of organisms to the cyclicity of environment is reflected in the wide spread occurrence of daily/circadian, lunar and annual/circannual cycles at almost all levels, sub-cellular, physiological and behavioral. Reproduction in wild vertebrates, subject to the seasonally changing environment, is thus a predominantly periodic phenomenon. As we go away from the polar region towards the tropics the seasons are less and less due to temperature and more and more due to wetness and dryness. But our understanding of how organisms have coped with this environmental challenge is amazingly poor.

Tropics, the richests on earth in faunal and floral regions, harbour not only maximum number and varieties of plants and animals but also half the world's population attesting to the diversity and favourable conditions that exist here for supporting life. Birds with the acquisition of homeothermy and ability to fly, thus having invaded every possible ecological niches, are ideal tools for the understanding of adaptational processes. A great deal of

This paper is dedicated to Prof. J.P. Thapliyal on his retirement from Banaras Kindu University, Varanasi

information is available on the avifauna of the tropics and, their
distribution, habit and habitat, and often nesting, have been
extensively described by naturalists. However, information on their
ecological relationships is meagre which is understandable keeping
in view of the extremely complex ecology of tropical environments.
Avian breeding cycles have been a topic of considerable interest
(Marshall, 1961; Murton and Westwood, 1977; Farner and Follett,
1980) and our current understanding of the phenomenon in the
tropics may be summarised as follows:
1. Reproductive seasons are spread out over the entire year in
keeping with the extended favourable conditions existent in tropics
but breeding in individual species is essentially periodic (Baker,
1938a, 1938b; Misra, 1948; Mehrotra, 1954; Thapliyal, 1954; Marshall,
1960; Miller, 1960) even in the fairly uniform tropical rain forests
(Fogden, 1972). The only truly arrythmic patterns have been reported
in some arid zone special to the central Australian deserts and
southwest Africa (Keast and Marshall, 1954; Keast, 1959; Immelmann,
1963).
2. Reproduction usually bears a close relationship with aperiodic
(Australian and African deserts) or periodic rainfall (Baker, 1938b;
Marshall, 1960). Thus birds in tropical rain forests breed in somewhat drier months. Although expectations are available, it is
generally believed that inhabitants of regions with one rainy and
one dry season breed during the rainy season whereas in areas with
two dry and two wet seasons breeding occurs either only during one
or during both rainy seasons (Immelmann, 1971; Gwinner, 1981).
3. Rainfall has been suggested as the environmental factor regulating
sexual cycles in tropical birds (Baker, 1938b; Lack, 1950, 1960;
Marshall, 1961) with different species responding to different
"monsoon factors" (Baker, 1938b; Marshall and Disney, 1957; Marshall,
1960, 1961; Lofts, 1975; Thapliyal, 1978, 1981). The only experimental evidence of dependence on rainfall however comes from the
arrythmic breeders of arid zone (Priedkalns.and Bennett, 1978).
4. An "internal rhythm" of reproduction was also suggested to govern
sexual periodicity in tropics (Misra, 1948, 1962; Thapliyal, 1954;
Marshall and Serventy, 1958). Distinct endogeneity of reproduction
as in many associated phenomena, has been experimentally demonstrated
in one of the tropical birds recently (Chandola et al., 1980, 1982a,
1982b).
5. The role of photoperiod as an environmental cue was totally ruled
out (Miller, 1959a; Marshall, 1960, 1961; Immelmann, 1971) despite
the fact that irrespective or tropical or temperate distribution
all bird species examined so far were found to be photosensitive
(Rollo and Domm, 1943; Marshall and Disney, 1956; Miller, 1959b,
1965; Disney et al., 1961; Lofts, 1962; Thapliyal and Tewary, 1964;
Epple et al., 1971; Chandola et al., 1973; Lewis et al., 1974;
Chandola and Thapliyal, 1977).

However, with carefully controlled laboratory experiments the
photoperiodicity of reproduction has been clearly established in

at least one tropical bird (Singh and Chandola, 1981; Chandola and Chakravorty, 1982; Singh and Chandola, 1982a, 1982b; Chandola et al., 1982c, 1982d).

From the most recent reviews available on the subject (Immelmann, 1971; Murton and Westwood, 1977; Farner and Follett, 1980; see specially Gwinner, 1981) the typical strategy suggested for the tropics seems to be that for predicitng the most favourable conditions for reproduction, birds make use of environmental cues ("proximate factors" e.g. rainfall) closely related or identical with the environmental factors most critical for reproduction (and hence of selective value -"ultimate factors". e.g. food supply, nesting sites). It is ironic that despite the recognition that reproduction in tropics is an essentially periodic phenomenon the only concrete strategy recognized for the tropics should be based on the only truly aperiodic breeders!

The aim of this review is to show, on the basis of the wealth of data provided by naturalists in the Indian subcontinent (Ali, 1949, 1979; Ali and Ripley, 1968-1974; Ali and Fatehally, 1967; Sinhji and Lavkumar, 1972; Fleming et al., 1979; Barnes, 1981), supported by laboratory experiments, that in keeping with the characteristic diversity of tropics a variety of strategies have been adopted for timing reproduction viz.:
1. Ultimate factors govering reproduction may be varied e.g. food nesting sites.
2. Proximate factors (e.g. food, rainfall) may be identical or closely related with the ultimate factors (e.g. food).
3. Proximate factors (e.g. daylength) may be qualitatively different from ultimate factors and may be temporarily closely or distantly related with ultimate factor (food). In the latter, supplementary factors (which for other species may be proximate factors e.g. rainfall) may be important.
4. The proximate factor may be an absolute prerequisite for the reproductive cycle or may be used only to synchronize an endogenous circannual rhythm with the most propitious time of the year.

THE INDIAN SUBCONTINENT

The Indian subcontinent constitutes a major part in the oriental realm with wide latitudinal (08°04' to 37°60'N) and longitudinal (68°07' to 97°25'E) extents. It's varied topography, great degree of insolation and a monsoon climate impart to it enormous complexity and habitat diversity. Ranging from the highest peaks of the earth in the north to the virgin tropical rain forests in the south, from the vast mangrove swamps of the east to the sprawling deserts of the west, home of some 1761 resident species of birds, it renders possible the study of reproduction in a wide variety of avifauna

in similar habitats and of similar/same species in diverse habitats
thus providing an excellent opportunity for simultaneous analysis
of various ecological factors operating. The tropic of Cancer roughly
divides the subcontintent into the subtropical (continental) and
tropical (peninsular) India. But because of the Himalayan massif in
the north the tropical zone (temperature-wise) extends well beyond
the tropic of Cancer. In winter the average temperature in the North
is 16°C and that in the peninsular India 25°C but the daily/seasonal
temperature differences in the North are great. Rainfall throughout
the subcontinent is received from the summer (Southwest, SW) and
winter (Southeast, SE) monsoon and imparts a seasonal diversity
especially in the north where six ecologically distinct seasons
exist- recognized as early as 3,000 years B.C. in Vedic literature
followed by a treatise by Kalidasa (500 years B.C.). Monsoon (S.W.)
arrives in penninsular India earlier (June) and leaves later
(September/October) than in the Northern parts (July to September).
This marked seasonality and distribution of rainfall is reflected
in natural vegetation of the land. Thus the western coast in the
South-West, Assam in the North-East and some slopes of the Western
Himalayan range are the areas of continuous rainfall and can support
evergreen rain forests. Some areas in southern peninsular receiving
low rainfall during SE monsoon exhibit two dry seasons. Areas with
marked dry seasons with heavy rainfall support monsoon (deciduous)
forests and savanna if rainfall is low (north and peninsula lowlands).
Savanna, scrub or even deserts are found where rainfall is too low
or too seasonal (the arid Rajasthan and Thar desert in North-West
and Deccan plateau in peninsula). However it must be born in mind
that due to extensive deforestation forests have given way to
savanna and cultivated land - only 20% of India being under forest
cover thus drastically affecting the present day wildlife.

PHENOMENON OF SEASONALITY IN REPRODUCTION

        Successful survival of a species is a measure of its ability
to breed at the time of the year most propitious for the upbringing
of the offspring. Favourable environmental conditions constitute
optimal temperature and photoperiod availability of nest sites/
material, and adequate food supply, for the growth of the hatchlings.
Of obvious selective value would be protection from predation.
There are two ways in which environmental factors can govern the
breeding season (Baker, 1938a):
- by exercising selection pressure in the course of evolution to
restrict breeding to a particular time of the year; environmental
factors in this category have been called ultimate factors (Baker,
1938a; Thomson, 1950), and may include all environmental factors
mentioned above; so far, however, food has been assumed to be the
most critical ultimate factor (Lack, 1950).
- by acting directly as a stimulus to initiate breeding process
physiologically; environmental factors of this kind are designated

as proximate factors (Baker, 1938a; Thomson, 1950). In view of the restriction imposed by the ultimate factors organisms will develop reproduction strategies permitting the use of the proximate factors for predicting the oncoming season.

While field workers, with their enormous data have somewhat facilitated the task of the identification of ultimate factors - by no means an easy effort - the identification of proximate factors is much more difficult. Not only because of paucity of data but also because of the technicalities involved. "The breeding season" in the eyes of the naturalists is characterized generally by the overt expressions of reproduction viz. nest building, courtship and egg-laying whereas the actual reproductive preparation invariably begins much earlier the overt events marking the culmination of reproductive phenomenon. Since proximate factors are causative factors that initiate the reproductive process (or entrain it to the ultimate factors) for their identification, in the first place information must be available on the exact time of the onset of breeding - when the sexually quiescent individual begins to be transformed into a sexually active one. This necessitates laboratory aid. Physiologically reproductive activity involves the growth and development of the primary sex organs (testis/ovary) the function of which is dual-production of gametes and secretion of sex hormones which permit growth/development of accessory sex-organs (gonoducts and accessory sex glands etc.) and secondary sex characters (beak pigmentation, plumage etc.) and influence/cause sexual behavior. Like in mammals, in birds also the external stimuli affect the gonads through a well developed neuroendocrine system.

FIELD STUDIES

Seventy birds were chosen at random, assigned to their respective orders and grouped accordingly to their breeding periods in calendar months (Fig. 1). The breeding periods were examined in relation to their distribution, feeding habit and habitat. Even among these few birds generalization strictly according to one criteria viz. habit and habitat or distribution, were not possible. However, certain trends become clear. The most characteristic feature was that despite some breeding going on in some part or the other throughout the year, irrespective of habit, habitat or distribution birds exhibit a clear cut periodicity in reproduction. There are a numbered few species, however, which appear to breed throughout the year although from the field data it is not known whether they are truly arrythmic or exhibit asynchronous populations.

It is significant however, that none of the species listed (Common Sand Grouse, Blue Rock Pigeon, Doves, Black-bellied Finch Lark, White throated munia) (Fig. 1, Groups 1 and 4) are birds of tropical rain forests where environmental conditions are more or

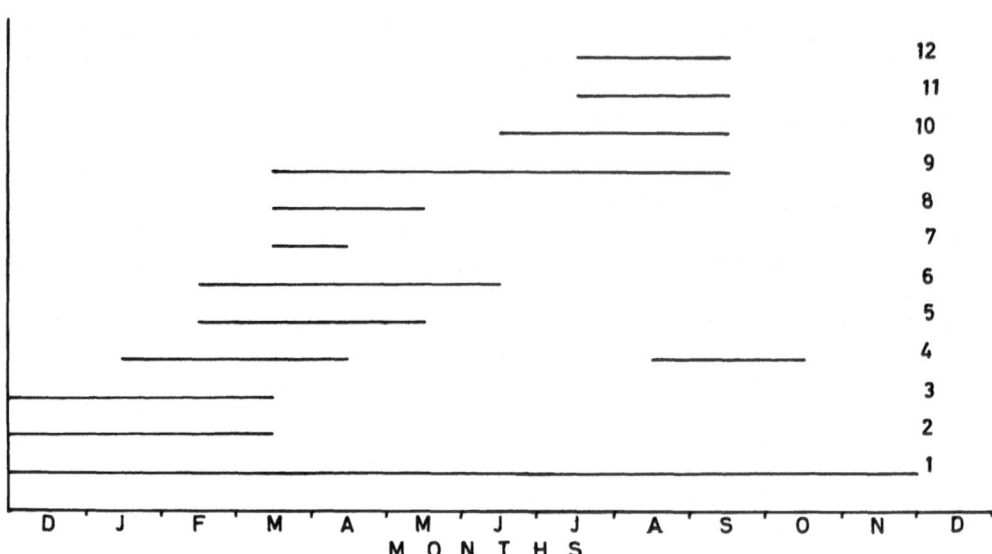

M O N T H S

Fig. 1. Breeding seasons based on about 170 Indian birds from
naturalists' data cited in the text. Ninenty five representative species are being mentioned in groups 1 to 12.
1. Black-bellied Finch Lark, Common Sand Grouse, Plain
Brown Munia, Blue Rock Pigeon, Red Turtle, Little Brown
and Ring Doves (overall distribution).
2. Slaty-headed Scimitar Babbler, Greyfronted Green Pigeon,
Southern Trogon, Heart Spotted Woodpecker, Pond Heron,
Cattle Egret, Little Cormorant, Darter, Spoonbill, little
Egret, White Ibis, Painted Stork (peninsular forms).
3. Laggar Falcon, Common Serpent Eagle, Ring tailed Sea
Eagle, Brahminy Kite, Dusky Horned Owl, White-backed
Vulture, Sparrow Hawk (overall distribution).
4. Dusky Crag Martin, Indian Cliff Swallow, Palm Swift,
Blue Rock Pigeon, Little Brown Dove, House Sparrow
(overall).
5. Himalayan Black Bulbul, White Cheeked Bulbul, Himalayan
Whisling Thrush, Hill Myna, Himalayan Gold Finch, Velvet
Fronted Nuthatch, Malabar Trogon, Wynaad Laughing Thrush,
Orange Minivet (Himalayan and Peninsular Hills).
6. Pied King Fisher, Common Indian King Fisher, White-
Breasted King Fisher, Common Indian Bee-eater, Small
Green Bee-eater (overall).
7. Stone Plover, Indian Ringed Plover, Large Swallow Plover,
Black-billed Tern, Larger River Tern (North Indian
Riverine).
8. Yellow Fronted Woodpecker, Small yellownaped Woodpecker,
Indian Hoopoe, White-eyed Buzzard, Indian Roller (overall).
9. Grey Partridge, White spotted Fantail Flycatcher, Common
Indian Nightjar, Black Drongo, White-eyed Tit, Singing

less stable throughout the year. Common sand grouse is exclusive to semi-arid/arid zone in the North as well as peninsular India. Same is true for Black-bellied Finch Lark.

It would be interesting to find out whether the continuous breeding in Sand Grouse is an adaptation to desert condition like in some desert birds of Australia and Africa (Immelmann, 1971) or is merely indicative of the reproductive potential shown by the o other columbiformes. Semiferal Blue rock pigeons, Doves and House sparrow are more or less continuous or biannual breeders. They perhaps reflect the more or less constant food supply since they are assured of living in close commensalism with man. The other commensals of man viz. crows, and myna however are strictly annual breeders. Unlike pigeons and sparrows they do not nest in house crevices and rafters of buildings thus perhaps more exposed to predation and competition for reproductive success. Protection from predation for reproductive success. Protection from predation may also be associated with increased reproductive potential (biannual) in some swifts and swallows (Fig. 1) which are known to build mud- cemented nests in crevices of houses and buildings or on inaccessible rocky cliffs often overhanging water.

Among the annual breeders it is clear that different birds prefer different periods for reproduction. It is difficult to categorize groups but certain trends become clear. In the lowlands (monsoon forest and savanna) in the North and throughout the peninsular India heavy SW monsoon (June to September) is the most favoured period (Fig. 1, Groups 9 to 12). Birds of lowland monsoon forests and open country breed during spring also (Fig. 1, Groups 6 to 9). In the Himalayan range (Fig. 1, Group 5) late spring and early summer are the peak periods irrespective of the habit. The onset of winter is obviously the critical factor. Similarly, the birds in the tropical rain forests of Western Ghats in the South

---

Bush Lark, Indian Grey Shrike, Sykes Tree Warbler, Purple Honey Sucker, White-winged Iora, Common Myna, Rain Quail (overall).

10. Long Leg Stilt, Glossy Ibis, Grey Duck, Sarus Crane, Pheasant-tailed Jacana, Little Cormorant, Darter, Cattle Egret, Little Egret, Little Grebe, Blue Heron, Spoonbill, White Ibis, Painted Stork (North Indian forms, Sarus Crane is exclusive to NI. Note that peninsular forms of Jacana sp. breed only during SW monsoon).

11. Black-headed Munia, Red Munia, Spotted Munia, Common Weaver Bird (overall except the peninsular Black-headed Munia).

12. Common Peacock, Jungle Bush Quail, Large Button Quail, Indian Bustard, Small Crown-crest Lark, Painted Partridge.

tend to breed in early spring which is the least humid period (Fig. 1, Groups 2 and 5). Extremely few annual breeders lay eggs during the post monsoon (early winter November-December) in the North whereas in some Southern parts receiving SE showers breeding is preferred during November to January (Fig. 1, Group 2, water birds). The large raptors as a rule (with the exception of Himalayan races which breed later) breed in later winter/early spring when visibility is clear and also young birds are plentiful and easy prey (Fig. 1, Group 3). Thus in this case food (ultimate factor) itself may act as a proximate factor to initiate breeding processes. The young of some larger forms e.g. Vultures and Eagles hatch out by late March. The wintering birds migrating to their breeding grounds in March/ April would also add to the food supply of young raptors which are known to have astonishingly voracious appetites and need to be contrinuously fed. Of special interest are the Owls which are highly specialized in the type of their prey. While the larger forest Eagle owls hunt for birds, barn and grass owls take small mammals and fish owl takes fish and the spotted owlet insects. While all others are early breeders the chiefly insect eating owlets breed later in spring alongwith a great number of other insectivorous birds. Thapliyal (1954) found a barn-owl-species (*Tytoalba*) to bread in monsoon Obviously detailed information on all these species will prove rewarding.

On the calendar scale soon after the raptors a great variety of species commence their nesting operations with the advent of Spring. It is significant that almost all of them chiefly feed upon insects which register an increase at this time associated with the prevernal and vernal flowering specially in Northern India. When nesting sites of birds in Fig. 1 were examined an almost common feature found was that they hardly apend any time building intricate nests perhaps in view of the restricted breeding time at their disposal. Availability of nesting site may be a crucial ultimate factor for a majority of Spring breeders since so far as food supply is concerned it is available in plenty in monsoon also, when a great number of other insectivorous breed. In some water birds which breed during Spring (e.g. Terns and Plovers) the availability of nesting sites may be of crucial importance in determining the time of reproduction. These birds lay their eggs in open on sandbanks in the larger rivers in the North and all reproductive activities must be finished before the rivers swell in summer due to melting of the Himalayan snows. Similarly breeding before the onset of monsoon may be an adaptation in Bee-eaters and King fishers alike which dig tunnel like nests in the banks of streams or ditches. The same may hold true for the species that nest in tree holes (Roller, Woodpeckers, Hoopoe) irrespective of distribution. What serves as the stimulus for breeding in birds? From the reports available they are never found to breed before February/March. The increasing daylength of Spring thus may be a likely candidate.

Seasonality in the subcontinent is predominantly due to rainfall and it is understandable that a great number of the birds breed during or immediately before monsoon when all kinds of food supply for carnivores, insectivores, herbivores, omnivores and grainivores alike is available in plenty. Some of these birds indicate a fair amount of flexibility in breeding in that nesting and egg-laying is definitely delayed in the year of a late rainfall (e.g. Sarus Crane, Weavers, Stewart's Wren Warblers, Little Grebe etc.). Same species breed a few weeks earlier or later at different latitudes depending upon the arrival of monsoon. This indicates that rainfall for these species might act as a proximate factor.

The most conspicuous of monsoon breeders are the water birds that congregate often in vast numbers, on India's lakes, artificial reservoirs and waterholes. Aquatic food is available in their habitual places but distribution all over during breeding not only increases chances of better food supply for the young ones but also lessens crowding and competitions for nesting sites. Herons, Egrets, Storks, Spoonbills and Ibises, Cormorants and Darters generally nest in colonies in trees and quite often several species will nest together. Sarus Cranes lay their eggs in reed on marshy ground and often in the centres of ponds. Grebes and Coots build floating nests on reeds protected all around by water. Jacanas also nest on the floating plants among which they feed. It is obvious that in all these cases (nests floating or in marsh) apart from food supply availability of nest sites which is closely associated with rainfall, too, may constitute an ultimate factor. Rainfall may thus provide the proximate stimulus for birds to fly to the breeding areas as also to trigger the neuroendocrine system initiating reproductive preparation. Dependence of breeding on proximate factors closely associated with the ultimate factors is a strategy characteristic of some opportunistic breeders also which are found breeding any time of the year (e.g. birds in central and Western Australian deserts and the temperate zone crossbill (Gwinner, 1981). Since soon after the summer monsoon the water levels of the ponds will drop and waterholes shrink and the birds disperse back to their feeding areas, it would be interesting to know how these birds have coped with the restriction on their breeding time - by a shortening of the progressive phase of the gonadal cycle or that of the incubation/gestation period, or by using rainfall only as a supplementary factor as may be the case with Weaver bird (see next section). In years of drought some of these waterbirds especially Sarus Crane have been known to fly long distances to some of the longer permanent lakes or rivers for breeding. This would indicate that the "urge to breed" was already there before the arrival of monsoon.

Among monsoon breeders a number of species are grainivorous. Some species viz. Spotted Munia, Black-headed Munia, Lal Munia and Weaver bird have been studied for their breeding cycles (Tewary and Thapliyal, 1963; Thapliyal and Saxena, 1964; Thapliyal and Pandha, 1965; Chandola et al., 1973). The Spotted Munia is distributed all over the Indian subcontinent. The gonadal development in this bird as in all munias coincides with the first monsoon showers. These birds depend exclusively on grass seeds and crop-grains and it is significant that compared to other monsoon breeders they should reproduce later with the availability of specific food.

LABORATORY STUDIES

        From the preceding section it is apparent that breeding at different periods with the diversity of seasons for the appropriate timing of reproduction birds have a variety of clues to choose from. For the timing proximate factor directly related to food supply (rainfall), food supply itself and proximate factors qualitatively different (daylength) but temporally close or distant were identified in the preceding section. These possibilities are now examined in the light of the laboratory data available. Gonadal cycles of a number of field-caught male birds were studied in the laboratory by Misra and his associates at Varanasi (Lat. 25°18'N) (Mehrotra, 1940, 1954; see reviews by Misra, 1962). Later on Thapliyal and his associates established reproductive cycles in captive birds by monthly laparotomy and *in situ* measurement of the gonadal size (Tewary and Thapliyal, 1963; Thapliyal and Saxena, 1964; Thapliyal and Pandha, 1965; Chandola et al., 1973). On comparison breeding activity appears to commence much earlier than that described on the basis of nesting/egg-laying (Fig. 1).

Food as a proximate factor

        Food supply (the ultimate factor) itself could act as a proximate factor where favourable food conditions are extended lasting long enough to enable the completion of all stages of reproduction and upringing of the young ones. Thus in the raptors the increased carnivorous diet which becomes available by October/November (the young of almost all monsoon breeders having hatched out) could afford a stimulus by itself. The fact that they invariably have very short breeding periods reflecting the short preparatory phase of the gonadal cycle, as observed in the laboratory, supports the possibility.

Rainfall as a proximate (and supplementary) factor

        Although a great number and varieties of birds are monsoon breeders and rainfall appears to be the proximate factor supporting experimental evidence is yet to come. Examination of gonadal cycles studied in the laboratory reveal that some birds (e.g. Spotted, Black headed and Red Munias) enter reproduction with the onset of monsoon whereas periods of egg-laying and nesting in

a majority of birds (e.g. Common Babbler, Redvented Bulbul, Cattle
Egret and Black Drongo) coincide with the first monsoon showers.
In the former rainfall appears to provide the stimulus some of
these birds (e.g. Spotted Munia) however, can be made to breed in
spring or winter irrespective of monsoon season (Chandola et al.,
1982a, 1982b) indicating that rainfall in this case, if at all, may
be used only as a supplementary information. In the latter obviously rainfall does not act as a proximate stimulus.

        Although rainfall has an excellent predictive value it is not
a reliable factor. There is evidence indicating that some monsoon
breeders (weaver bird) have indeed recognized this unreliability of
the onset of monsoon and use it not as a cue for reproduction
preparation but perhaps as a supplementary factor to bring reproduction to its finale by triggering the processes of nest building/
egg-laying. It would definitely be more advantageous for the bird
to be almost prepared for breeding activities on the arrival of
monsoon - especially if there is some kind of restriction imposed
on the termination of breeding thus requiring quick gonadal growth
(e.g. the young of Egrets, Herons, Cormorants, Darters etc. must
fledge out before the shrinking up of the waterholes and lakes in
the breeding grounds). The gonads of cattle Egret begin to develop
much earlier that the onset of rainfall at Varanasi (25°N, Mehrotra,
1954). In fact a complete dependence of the reproductive urge on
monsoon would limit the chances of these strong fliers to venture
new breeding grounds. In the years of drought Sarus cranes emigrate
to larger lakes and rivers where the conditions for breeding are
favourable (Sinhji and Lavkumar, 1972).

Daylength as a proximate factor

        Since the amplitude of the annual photocycle decreases with
latitude it has generally been believed that daylength might act
serve as an effective cue for timing reproduction in tropics (Miller,
1959a; Marshall, 1960, 1961; Immelmann, 1971). However, a great
number of birds do come into breeding during spring and early summer
some even continuing into monsoon (Fig. 1, Group 9). While the food
supply of spring might act as a proximate factor for these latter
it is inconceivable that the same may hold true for the earlier
breeders in view of the restriction imposed upon them by the impend-
ing SW monsoon in June and thus damaged nests. Some of the spring
breeders (Myna, Bulbul) have been shown to respond to photoperiod
exactly like in temperate zone birds (25°N; Thapliyal, 1981). Our
studies have clearly established that in at least two species
(distributed all over subcontinent) the daylength might serve as
a cue for reproduction although it may be utilized in different
ways - as a "driver" or synchronizer of an endogenous rhythm.

Daylength as a driver: Weaver bird

        Nesting and egg-laying in weaver bird begin soon after first
monsoon (SW) showers depending on the area (Fig. 1) as evident from
Fig. 2, gonads begin to develop in Spring, reach a miximum in June
and rapidly regress thereafter. The adult males wear a bright yellow
nuptial plumage while the females are dull henny. The onset of re-
rpoduction can be simulated in artificial conditions in that
initiation of gonadal development occurs in 11-12h daily photoperiod
(spring daylength) and maximal gonadal activity is attained in 13h
(summer daylength) or longer daily photoperiod (Fig. 3). Short day-
lengths completely suppress the reproductive cycle (as judged by
yellow plumage and weight and histology of gonads, Thapliyal and
Saxena, 1964). This indicates a definite photoperiodic requirement
for the onset of breeding in this bird. Long photoperiods abolish
the reproductive cycle by maintaining the gonads indefinitely active
(Thapliyal and Saxena, 1964; Thapliyal and Tewary, 1964; Chandola
and Thapliyal, 1978; Singh and Chandola, 1981). These findings
suggest that the photoperiod is a prerequisite for gonadal growth
in this bird. Thus photoperiod appears to act as a driving agent
(Chandola et al., 1982a, 1982c).

        Use of the daylength as a cue for seasonal reproduction would
need some mechanism to measure the length of the photoperiod.
Distinct endogenous circadian rhythms of photosensitivity have been
demonstrated in ahemeral and asymmetrical skeleton photoperiods
(Chandola et al., 1976b; Singh and Chandola, 1982b) as in several
temperate zone species (Follett, 1973; Farner and Follett, 1980).
Increasing daylength of spring thus leads to gonadal development as
a result of an interaction with circadian components. Thus photoperiod in weaver bird as in many other monsoon breeders may predict
the most favourable season well in advance of the reproductive
climax i.e. nest building and egg-laying, thus allowing sufficient
time for the growth of the gonads.

Daylength as a synchronizer: Spotted munia

        Spotted munia is a monsoon (SW) breeder which feeds mainly
upon grass seeds and crop-grains. In nature as well as in captivity
the gonads of Spotted munia begin to develop from June to July,
attain peak of activity by September/October and regression occurs
from October to November. Thereafter gonads remain quiescent from
December to June (Thapliyal and Pandha, 1965; Chandola and Thapliyal,
1978; Fig. 4). In this bird customary long and short daylengths
have no immediate effects on the gonadal development (Chandola et
al., 1975). In constant darkness (Chandola et al., 1976a) or
continuous illumination (LL) the reproductive cycle freeruns with
a period of about 10 months (Chandola et al., 1982a, 1982b; Fig. 4),
thus exhibiting endogenous circannual nature. Circannual cycles
have been demonstrated in a number of physiological and behavioral

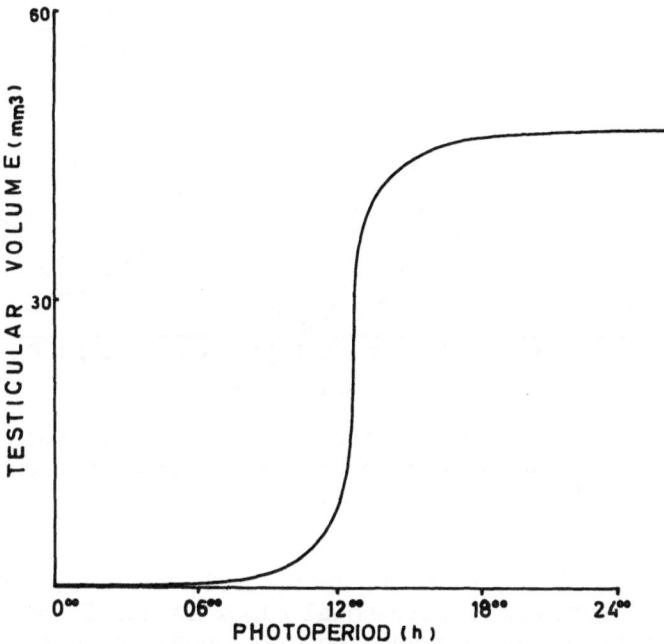

Fig. 2. Photoperiodic response-threshold of weaver bird at 25°N.
Based on Singh and Chandola, 1981.

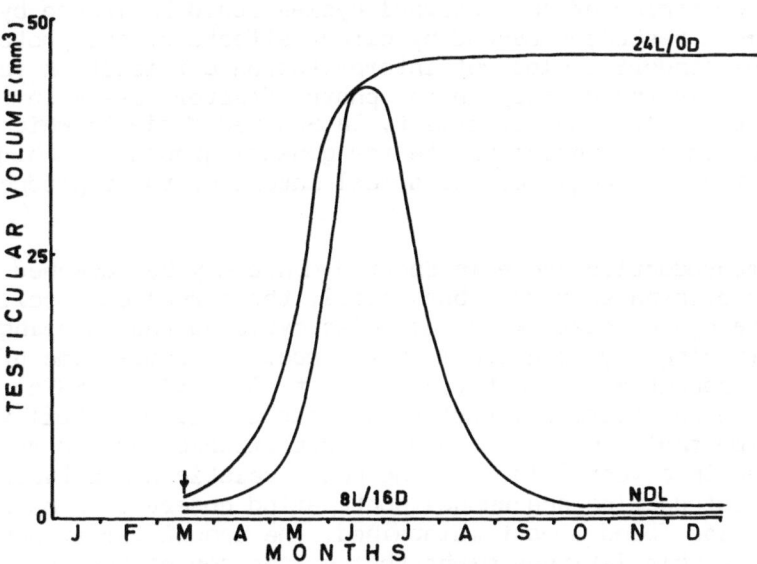

Fig. 3. Effect of continuous illumination (24L/0D) and short days
(8L/16D) on testicular cycle of weaver bird. Based on
Singh and Chandola, 1981.

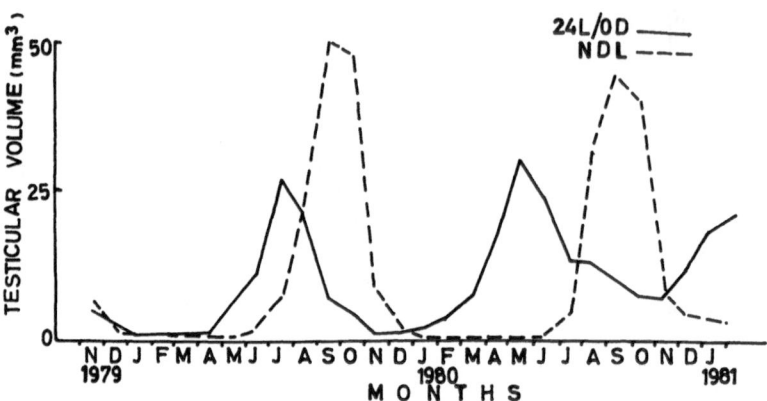

Fig. 4. Circannual rhythm of reproduction in adult Spotted munia
(*Lonchura punctulata*) in continuous illumination (24L/0D).
Based on Chandola et al., 1982c.

functions (Lofts, 1964; Gwinner, 1975, 1981; Berthold, 1978) but
the concept of autonomous circannual clock is still open to cri-
ticism. Majority of the evidence on circannual rhythms comes from
birds held in constant light/dark (LD) cycles which may not
necessarily be an information free environment (Hamner, 1971; Sansum and King, 1976; Farner and Wingfield, 1978). Also, all these
species are strongly photosensitive in which any endogenous contribution to the timing of the seasonal cycles could be masked by the
physiological responses evoked by direct effects of the photoperiod
per se thus rendering clearcut interpretation difficult. A further
complication is the development of photorefractory state in these
birds following long-day treatment. In Spotted Munia in which.
photoperiod is not a prerequisite for gonadal growth no ambiguity
exists as to the endogenous circannual nature of the reproductive
cycle.

The reproductive cycle in Spotted munia may be governed by
endogenous circannual rhythm but, still, the circannual oscillator
needs to be synchronized with the solar cycle so that upbringing
of the offspring may coincide with the most favorable time of the
year. In a recent experiment (Chandola et al., 1982a, 1982c) birds
were held in an artificial photocycle with a period of half year
only (T 6-months). In this schedule birds experienced two reproductive cycles in a year (Fig. 5). The phase relatioship between the
cycle of reproduction and annual photoperiod observed in natural
conditions was found almost maintained. The annual variations in
daylength at this latitude might thus act as one of the synchro-
nizers of the circannual reproductive cycle of Spotted munia. There
is evidence indicating that apart from daylength seasonal decline
in food intake/availabitlity may be associated with the controlling

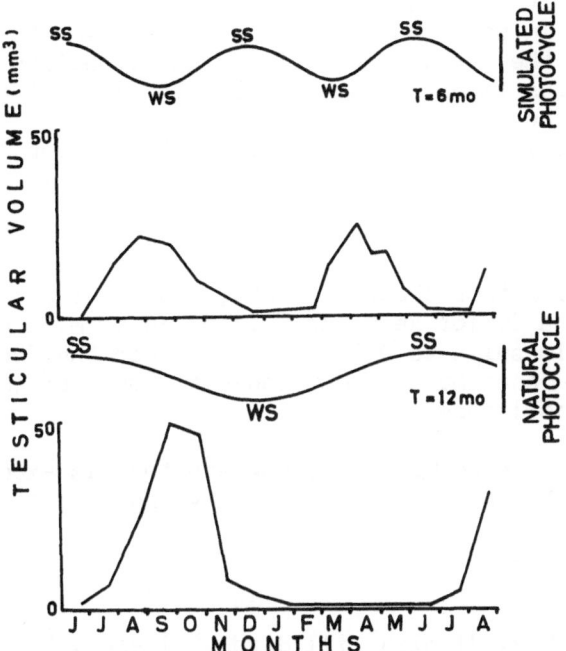

Fig. 5. Photoperiodic synchronization of the reproductive cycle of
Spotted munia (*Lonchura punctulata*) in an artificially
simulated six-month photocycle. Based on Chandola et al.,
1982a.

mechanism(s) involved in the seasonal reproduction of Spotted munia
(Chandola et al., 1982a).

In the Indian subcontinent monsoon no doubt is the seasonal
abiotic factor heralding propitious times which last long enough
for the termination of all reproductive processes and provide ample
food for the upbringing of the offspring. It is periodic and has
an excellent predictive value (and hence an ideal choice as a
proximate factor for some monsoon breeders) but equally characteristic is its unreliability in that it may arrive a few weeks
earlier or later. The best strategies therefore would be those which
recognize and take into account the *circa* nature of the unreliability
so that not depending directly on the monsoon as a clue the birds
can still recognize it as the most favorable period. Hence the use
of a totally reliable temporally related factor (photoperiod) for
initiating reproductive processes using the rainfall as the final
trigger-like in Weaver bird. The circannual rhythm as shown by
Spotted munia is another example of the recognition of the subtle
year to year variations in the seasons. While the circannual rhythm
has been shown to be synchronized by changing daylength with the

monsoon period it may also be "open" to adjustments by factors like
rainfall/food. In what manner, if at all, remains to be seen. The
recognition of the *circa* nature of the unreliability of rainfall
may be of obvious importance in areas with distinct and diverse
seasons as in the North India as also in birds with special food
preferences, as is invariably the case with monsoon breeders.

ACKNOWLEDGEMENTS

    The senior author thanks the UGC, India for a Career Develop-
ment Award and a Junior Research Fellowship to M.S. under the
Programme. Fellowships to D.B. and M.B. under the Eco-Development
Project of the Department of Environment, Govt. of India, are also
gratefully acknowledged.

    Thanks are due to Mr. P. C. Lakhera for his excellent technical
assistance and Mr. K. N. Nautiyal for typing the manuscript.

REFERENCES

Ali, S., ed., 1949, "Indian Hill Birds," 1 edn., Oxford University
    Press, Bombay, Calcutta, Madras.
Ali, S., ed., 1979, "The Book of Indian Birds," 11th edn., Pub.
    Bomb. Nat. Hist. Soc., Bombay.
Ali, S., and Fatehally, L., eds., 1967, "Common Birds," National
    Book Trust India.
Ali, S., and Ripley, D., eds., 1968-1974, "Hand Book of the birds
    of India and Pakistan," 10 vols, Oxford University Press,
    Bombay.
Baker, J. R., 1938a, The Evolution of breeding season, in:"Evolution:
    Essays on Aspects of Evolutionary Biology," G. R. de Beer, ed.,
    Oxford University Press, Oxford.
Baker, J. R., 1938b, The relation between latitude and breeding
    seasons in birds. Proceedings of the Zoological Society
    London.
Berthold, P., 1978, Concept of endogenous control of migration in
    warblers, in: "Animal Migration, Navigation and Homing," K.
    Schmidt Koenig and W. T. Ketton, eds., Springer Verlag, Berlin,
    Heidelberg, New York.
Barnes, H. E., ed., 1981, "The birds of India," Vol. I, II, 1st edn.,
    Cosmo Pub., New Delhi.
Chandola, A., Bhatt, D., and Pathak, V. K., 1982a, Environmental
    manipulation of the reproductive cycle of Spotted munia, in:
    "Avian Endocrinology: Environmental and Ecological Perspec-
    tives," S. Mikami et al., eds., Japan Sci. Soc. Press, Tokyo.
Chandola, A., and Chakravorty, K., 1982, Termination of seasonal
    breeding in Indian Weaver bird, J. Exp. Zool., 222:169-172.

Chandola, A., Pathak, V. K., and Bhatt, D., 1982b, Evidence for an
    endogenous circannual component in the control of annual gonadal
    cycle in Spotted munia, J. Interdiscipl. Cycle Res., 13:281-
    286.
Chandola, A., Pavgi, S., and Pathak, V. K., 1980, Environmental
    correlates of the annual fattening cycle of Spotted munia
    (Lonchura punctulata), J. Interdiscipl. Cycle Res., 2:125-133.
Chandola, A., Pavanaskar, J., and Thapliyal, J. P., 1975, Photo/
    Scotoperiodic responses of a sub-tropical finch, Spotted munia
    in relation to seasonal breeding cycle, J. Interdiciplin. Cycle
    Res., 6:189-202.
Chandola, A., Pavanaskar, J., and Thapliyal, J. P., 1976a, Decreasing
    photoperiods and the testicular cycle of Spotted munia, J.
    Interdiscipl. Cycle Res., 7:95-100.
Chandola, A., Singh, S., and Bhatt, D., 1982c, Interaction of Photo-
    period and circadian/circannual rhythms in the seasonal re-
    production of Indian birds, in: "Proceedings 9th International
    Symposium on Comp. Endocrinology," B. K. Lofts, ed., Hong-
    Kong University Press, Hong-Kong (in press).
Chandola, A., Singh, S., and Chakravorty, K., 1982d, Reproductive
    periodicity in the weaver bird: role of photoperiod and of
    gonadal hormones..28th Congress Internationalis ornithologicus
    Moscow (in press).
Chandola, A., Singh, R., and Thapliyal, J. P., 1976b, Evidence for a
    circadian oscillation in the gonadal response of tropical
    weaver bird to programmed photoperiods, Chronobiologia, 3:
    219-227.
Chandola, A., Thapliyal, J. P., and Murty, G. S. R. C., 1973, Photo-
    periodism and sexual activity in Indian birds. Proc. UNESCO
    Int. Cong., Paris "The sun in the service of mankind," B 14:
    1-10.
Chandola, A., and Thapliyal, J. P., 1977, Photoperiodic responses
    of Indian birds, in:"Avian Endocrinology," B. K. Follett, ed.,
    University College Northwales, Bangor.
Chandola, A., and Thapliyal, J. P., 1978, Regulation of reproductive
    cycles of tropical Spotted munia and Weaver bird, in: "Environ-
    mental Endocrinology," I. Assenmacher and D. S. Farner, eds.,
    Springer Verlag, Heidelberg, New York, Berlin.
Disney, H. J. deS., Lofts, B., and Marshall, A. J., 1961, An Experi-
    mental study of the internal rhythm of reproduction in the
    Red-billed Dioch Quelea quelea by means of photostimulation,
    with a note on melanism induced in captivity, Proc. Zool. Soc.
    London, 1369:123-129.
Epple, A., Orians, G. H., Farner, D. S., and Lewis, R. A., 1971,
    The photoperiodic testicular response of a tropical finch
    Zonotrichia capensis costaricensis, Condor, 74:1-4.
Farner, D. S., and Follett, B. K., 1980, Reproductive periodicity
    in Birds, in:"Hormones and Evolution," E. J. W. Barrington,
    ed., Academic Press, New York, San Fransisco, London.
Farner, D. S., and Wingfield, J. C., 1978, Environmental endocrino-

logy and the control of annual reproductive cycle in Passerine
    birds, in:"Environmental Endocrinology," D. S. Farner and
    I. Assenmacher, eds., Springer Verlag, Berlin, Heidelberg,
    New York.
Fleming, R. L., Sr., Fleming, R. L. Jr., and Bangdel, L. S., ed.,
    1979, "Birds of Nepal," 2nd edn., Avalok Publishers, Nepal.
Follett, B. K., 1973, Circadian rhythms and photoperiodic time
    measurement in birds, J. Reprod. Fertil., Suppl., 19:5-18.
Fogden, M. P., 1972, The seasonality and population dynamics of
    equatorial forest birds in Sarawak, Ibis, 114:307-343.
Gwinner, E., 1975, Circadian and circannual rhythms in birds, in:
    "Avian Biology," D. S. Farner and J. R. King, eds., vol. 5,
    Academic Press, New York, San Fransisco, London.
Gwinner, E., 1981, Circannual systems, in:"Handbook of Behavioral
    Neurobiology," J. Aschoff, ed., vol. 4, Plenum Publishing
    Corporation, New York.
Hamner, W. M., 1971, On seeking an alternative to the endogenous
    reproductive rhythm hypothesis in birds, in:"Biochronometry,"
    M. Menaker, ed., Nat. Acad. Sci., Washington, D.C.
Immelmann, K., 1963. Drought adaptation in Australian desert birds.
    Proceedings 12th International ornithological Congress.
Immelmann, K., 1971, Ecological aspects of periodic reproduction,
    in:"Avian Biology," D. S. Farner and J. R. King, eds., vol. 1,
    Academic Press, New York, San Fransisco, London.
Keast, A., 1959, Australian birds: Their zoogeography and adaptations
    to an arid environment, in:"Biogeography and Ecology in Austra-
    lia," A. Keast, R. L. Crocher, and C. S. Christian, eds., Junk
    Publishers, The Hague.
Keast, J. A., and Marshall, A. J., 1954, The influence of draught
    and rain-fall on reproduction in Australian desert birds,
    Proc. Zool. Soc. London, 124:493-499.
Lack, D., 1950, The breeding seasons of European birds, Ibis, 92:
    288-316.
Lack, D., 1960, The influence of weather on Passerine migration,
    Auk, 77:171-209.
Lofts, B., 1962, Photoperiod and the refractory period of reproduc-
    tion in an equatorial bird Quelea quelea, Ibis, 104:404-414.
Lofts, B., 1964, Evidence of an autonomous reproductive rhythm in
    an equatorial bird Quelea quelea, Nature, 201:523-524.
Lofts, B., 1975, Environmental control of reproduction, Symp. Zool.
    Soc. Lond., 35:177-197.
Lewis, R. A., Farner, D. S., and King, J. R., 1974, Photoperiodic
    responses of a subtropical population of the finch Zonotrichia
    capensis hyploleuca, Condor, 76:233-237.
Marshall, A. J., 1960, The Environment. Cyclical reproductive activity
    and behavious in birds, Symp. Zool. Soc. Lond., 2:53-67.
Marshall, A. J., 1961, Breeding seasons and migration, in:"Biology
    and Comperative Physiology of Birds," A. J. Marshall,ed., vol. 2,
    Academic Press, New York, London.

Marshall, A. J., and Disney, J. J. deS., 1956, Photostimulation of
    an equatorial bird (*Quelea quelea*), Nature, 177:143-144.
Marshall, A. J., and Disney, H. J. deS., 1957, Experimental induc-
    tion of the breeding season in a xerophilous bird, Nature, 180:
    647-649.
Marshall, A. J., and Serventy, D. L., 1958, The internal rhythm of
    reproduction in xerophilous birds under conditions of
    illumination and darkness, J. exp. Biol., 35:666-670.
Mehrotra, S. N., 1940, D. Sc. Thesis, BHU.
Mehrotra, S. N., 1954, Some thoughts on avian reproduction, Sci.
    Res. Banaras Hindu University, 5:1-29.
Miller, A. H., 1959a, Reproductive Cycles in an equatorial sparrow,
    Proc. Natl. Acad. Sci., 45:1095-1100.
Miller, A. H., 1959b, Response to experimental light increments by
    Andean sparrows from an equatorial area, Condor, 61:344-347.
Miller, A. H., 1960, Adaptation of breeding schedule to latitude.
    Proceedings of the 12th International ornithological Congress.
Miller, A. H., 1965, Capacity for photoperiodic response and endo-
    genous factors in the reproductive cycles of an equatorial
    sparrow, Proc. Natl. Acad. Sci., 54:97-101.
Misra, A. B., 1948, Presidential address. Proceedings Science
    Congress 35th session.
Misra, A. B., 1962, The recurrent sexual cycles of birds. Presidential
    address. Second All India Congress Zoology.
Murton, R. K., and Westwood, N. J., eds., 1977, "Avian Breeding
    Cycles," Clarendon Press, Oxford.
Priedkalns, J., and Bennett, R. K., 1978, Environmental factors
    regulating gonadal growth in the Zebra finch.*Taeniopygia
    guttatta castanotis*, General and Comparative Endocrinology,
    34:80.
Rollo, M., and Domm, L. V., 1943, Light requirements of the weaver
    finch. I. Light period and intensity, Auk, 60:357-367.
Singh, S., and Chandola, A., 1981, Photoperiodic control of seasonal
    reproduction in the tropical weaver bird, J. Exp. Zool., 216:
    293-298.
Singh, S., and Chandola, A., 1982a, Seasonal variation in photo-
    gonadal response of weaver bird, General and Comparative Endo-
    crinology, 48 (in press).
Singh, S., and Chandola, A., 1982b, Photoperiodic time measurement
    in seasonal reproduction of weaver bird, General and Comparative
    Endocrinology (in press).
Sansum, E. L., and King, J. R., 1976, Long term effects of constant
    photoperiod on testicular cycles of white crowned sparrow,
    Zonotrichia leucophrys gambelii, Physiol. Zool., 49:407-416.
Sinhji, R. S., and Lavkumar, K. S., eds., 1972, "Sixty Indian Birds,"
    1st edn., Pub. Govt. of India, New Delhi.
Thapliyal, J. P., 1954, Light and sexual cycle of owls, J. Sci. Res.
    Banaras Hindu University, 5:31-48.
Thapliyal, J. P., 1978, Reproduction in Indian birds, Pavo, 16:151-
    161.

Thapliyal, J. P., 1981, Endocrinology of avian reproduction. Proceed-
    ings 68th Symp. Science Congress, Part II.
Thapliyal, J. P., and Pandha, S. K., 1965, Thyroid-gonad relation-
    ship in Spotted munia *Uroloncha punctulata*, J. Exp. Zool., 158:
    253-259.
Thapliyal, J. P., and Saxena, R. N., 1964, Absence of refractory
    period in the common weaver bird, Condor, 66:199-208.
Thapliyal, J. P., and Tewary, P. D., 1964, Effect of light on the
    pituitary, Gonad and plumage pigmentation in the Avadavat
    (*Estrilda amandava*) and Baya Weaver (*Ploceus philippinus*),
    Proc. Zool. Soc. Lond., 142:67-71.
Tewary, P. D., and Thapliyal, J. P., 1963, Sexual cycle and secondary
    sex characters of the male lal munia *Estrilda amandava*, callule
    LXIII, 3:361-365.
Thomson, A. L., 1950, Factors determining the breeding seasons of
    birds: an introductory review, Ibis, 92:173-184.

NUTRITIONALLY RELATED METABOLIC ADAPTATIONS OF CARNIVORES AND RUMINANTS

James G. Morris and Quinton R. Rogers

Departments of Animal Science and Physiological Sciences
School of Veterinary Medicine
University of California, Davis, CA 95616, U.S.A.

INTRODUCTION

Among the present day mammals, ruminants and true carnivores represent two groups with widely divergent dietary habits and digestive systems. Both groups are presumed to have arisen from common ancestors -primitive insectivores which appeared during the Cretaceous period (Romer, 1966; Geist, 1972; Colbert, 1980). During the course of evolution, the metabolism of these mammals appears to have been adapted to accommodate the end products absorbed from their respective digestive systems. The purpose of this paper is to briefly review and contrast some of these metabolic adaptations and relate them to the nutrition of these two groups of mammals. Because of their economic importance, ruminants belonging to the family Bovidae (which includes cattle, sheep and goats) have been most extensively studied and will be used as type examples. However, living members of the sub-order Ruminantia comprise about 150 species with considerable diversity of feeding habits (Van Soest, 1982) and these examples may not truly represent all species.

In contrast to ruminants, there is a dearth of information on the metabolism of pure carnivores. Many animals classified as carnivores such as dogs, raccoons and bears have omnivorous dietary habits, and some (e.g., giant panda) are strict herbivores and hence are not included in this discussion of carnivores. As some data are available for studies on the domestic cat as a laboratory and companion animal, it will be used as the type example of a pure carnivore in this paper.

DIETS OF RUMINANTS AND CARNIVORES

Ruminant and carnivore diets are markedly different in digestive and chemical characteristics as shown in Table 1. Diets of grazing ruminants comprise mainly structural and photosynthetic parts of plants, high in cell wall components which are refractory to digestion by mammalian enzymes. The development of pregastric microbial fermentation in ruminants facilitated utilization of some of the energy in cell walls. This adaptation of the digestive tract has been credited with the increased number of ruminant species which coincided with the development of the Angiosperms as dominant plants. However, it may not have been the primary force. Predigestion detoxifies many secondary plant substances (James et al., 1975) which may be produced by plants as a defense against herbivory. Thus, a greater diversity of plant species and dietary choice opens the potential for wider adaptation to the pregastric fermenter than the simple stomached animal (Freeland and Janzen, 1974). Metabolism of certain plant compounds by ruminants leads to the production of xenobiotics, such as benzoic acid, which are not further oxidized. In ruminants, benzoic acid is conjugated with glycine and the resulting compound, hippuric acid, is excreted in urine. On the other hand, urine of carnivorous animals normally contains only low concentrations of hippuric acid. Cats are particularly intolerant to benzoic acid in their diet (Bedford and Clark, 1972). While the cat uses the glycine conjugation it is unable to utilize the alternate glucuronide conjugation (Hirom et al., 1977).

ADAPTATIONS ASSOCIATED WITH ENERGY AND PROTEIN METABOLISM

Although ruminant diets are characteristically high in carbohydrates, mainly hexose and pentose polymers, microbial fermentation of these substrates to volatile fatty acids results in little glucose per se being absorbed from the gut. Some starch may escape ruminal digestion, but the intestine of the adult sheep appears to have only a limited ability to transport glucose against a concentration gradient (Scharrer et al., 1976). Thus, for different reasons, the livers of carnivores and ruminants receive little glucose from the gut. The liver of the rat and many other non-ruminants has two enzymes that catalyze the phosphorylation of glucose to glucose-6-phosphate-glucokinase and hexokinase. The former enzyme which has a high K_m for glucose than hexokinase (therefore operating only during high concentration of glucose in the liver) is virtually absent from the liver of ruminants and cats though present in the liver of the dog, mouse and rat (Ballard, 1965). It has been suggested that the absence of glucokinase from the livers of ruminants and cats reflects the low direct contribution of dietary carbohydrate to blood glucose. The presence of glucokinase in the liver of the dog supports our view that the dog, though classified as a carnivore has a metabolism more closely

Table 1. Some characteristics of ruminant and carnivore diets

Characteristics	Ruminant diets	Carnivore diets
Nature of diet	Structural and photosynthetic parts of plants	Animal tissue
Digestibility	Cell wall components are refractory to mammalian enzymes	Readily digested by mammalian enzymes
Organic matter digestibility (%)	Most forages <65	>85
Presence of natural toxins	Secondary plant compounds	None in species normally eaten
Proximate contituents of the diet		
Carbohydrates	High	Low
Fat	Low	High
Protein	Low (generally)	High

allied to that of an omnivore. This view is also consistent with
the proposed phylogeny of the carnivora, in which the Canidae and
Felidae developed as separate families comparitively early in the
evolution of placental mammals, in the late Eocene period (Romer,
1966).

Both, adult ruminants and cats efficiently digest most starches
and sugars. However, adult ruminants are prone to digestive upsets
if large quantities of readily available carbohydrates are suddenly
introduced into the diet. On such occasions production of lactic
acid (both D and L isomers) by the rumen microflora can cause severe
metabolic acidosis and even death (Dunlop and Hammond, 1965). Conversion of L-lactate to pyruvate occurs in the cytosol of all tissues,
especially heart and skeletal muscle under the action of the NAD-
dependent L-lactate dehydrogenase. The analogous enzyme for D-lactate
appears to be absent in mammals and instead D-lactate is metabolized
by the mitochondrial FAD-containing enzyme D-2 hydroxy acid dehydrogenase (Giesecke and Stangassinger, 1980). Metabolism of D-lactate
proceeds at a slower rate than L-lactate as it has to pass the mitochondrial membrane and the D-2-hydroxy acid dehydrogenase is inhibited by pyruvate and oxalacetate, both of which would be produced
from the simultaneous metabolism of L-lactate.

Young ruminants before development of a functional rumen e.g.,
calf less than 2 months of age, have a limited capacity to digest
starch (Thivend et al., 1980); but this capacity can be increased
by progressive introduction of starch into the diet and by partial
hydrolysis of the starch. Presumably this enhanced ability to
utilize starch is achieved by induction of amolytic enzymes and
disaccharidases (Toofanian et al., 1973).

The normal glucose concentration of the blood of adult ruminants
(e.g., cow) is in the range of 2.5-4.2 mmolar which is much lower
than that of simple stomached herbivores (e.g., horse) which is in
the range of 4.2-6.4 mmolar (Kaneko, 1980). Glucose concentration
in the blood of the pre-ruminant calf approaches that of the simple
stomached herbivore but declines with age (McCandless and Dye, 1950).
It has been suggested that this decline coincides with the replace-
ment of fetal erythrocytes (high in glucose) with those of the adult
and development of volatile fatty acid production in the large
intestine (Vandersall et al., 1957; Liang et al., 1967).

Domestication of ruminants and selection for high levels of
milk production in cows and multiple births in ewes have resulted
in an increased incidence of diseases associated with hypoglycemia
and ketosis. Dairy cows in early lactation are frequently unable
to provide adequate glucose for both mammary gland synthesis
(lactose, amino acids and glycerol) and normal body functions.
Similarly, ewes carrying multiple fetuses in late pregnancy if
subjected to a reduction in feed intake can become hypoglycemic

due to the high and constant fetal demand for glucose.

Ruminants have adapted to low inputs of glucose from their digestive systems by conservation and by the maintenance of a high rate of gluconeogenesis. The main conservation mechanism utilized by ruminants is avoidance of the use of glucose for lipogenesis. At the major sites of lipogenesis in ruminants (adipose tissue and mammary gland) acetate rather than glucose is used as a substrate for fatty acid synthesis. Avoidance of use of glucose is achieved in these tissues by a lack of the cytosolic enzyme ATP-citrate lyase which prevents acetyl CoA generated in the mitochondrion from pyruvate moving to the cytosol (Bauman and Davis, 1975). Furthermore, ruminant adipose tissue and mammary glands lack NADP-malate dehydrogenase, so this source of reducing equivalents is not available for fatty acid synthesis. Instead ruminants rely on the pentose phosphate cycle and the NADP isocitrate cycle to generate NADPH.

Ruminants are characterized by having their gluconeogenic pathway more or less permanently switched on (Lindsay, 1978). Unlike non-ruminant species, the rate of gluconeogenesis in ruminants is greatest after a meal, when there is a large influx of metabolites from the digestive system. The rate slowly declines during starvation. The release of glucose from ruminant liver is also greated during the fed than the fasted state; whereas, in most non-ruminants the liver in the fed state has a net uptake of glucose. In ruminants, the volatile fatty acid propionate is the main gluconeogenic substrate for the liver; amino acids act as a carbon source only to a limited degree (Lindsay, 1982). Comparisons of the specific activities of precursor amino acids to circulating glucose suggests that only about 14% of the glucose output of the liver comes from amino acids. Alanine, glutamate/glutamine are the most significant precursors (Lindsay, 1982). The sheep liver takes up considerable quantities of glycine, but this may be a reflection of the utilization of glycine for hippuric acid synthesis.

Glucose homeostasis in the cat ingesting a diet of animal tissue has many parallelisms to those encountered by ruminants. Both groups of mammals absorb little glucose as an end product of digestion. However, the metabolites absorbed by the cat following assimilation of food are largely amino acids and fats. Direct transfer and storage of absorbed fat into adipose tissue circumvents involvement of significant quantities of glucose. Glucogenic amino acids then provide the main substrate for glucose production by liver. Most mammals have the capacity to adapt to various levels of protein in the diet (Harper, 1965; Knox and Greengard, 1965). When the diet contains a suboptimal amount of protein, the activities of the amino acid catabolic enzymes decrease, conserving protein. Similarly, when a high protein diet is consumed, the activities of the catabolic enzymes increase facilitating utilization of the excess amino acids. Activities of the enzymes of the urea cycle follow a similar pattern allowing disposal of the nitrogen load (Schimke, 1962).

Studies on the effect of protein level in the diet on the activity of transaminases and urea cycle enzymes have shown that the adult cat, ulike most other mammals, does not regulate the activities of these enzymes (Rogers et al., 1977; Rogers and Morris, 1980). The adult cat maintains these enzymes at a high level at all times, hence is unable to conserve nitrogen to the same degree as other mammals such as the rat and sheep. This adaptation is advantageous to a carnivore which always consumes a high protein diet (animal tissue) as it allows disposal of excess amino acid nitrogen without a time lag for enzyme induction. Presumably the cat in its evolutionary history has maintained a strict carnivorous diet, and it has not been exposed to selection pressures for nitrogen conservation.

In simple stomached animals, such as the rat, starvation or treatment with glucocorticoids increases the activity of the gluconeogenic enzyme phosphoenolpyruvate carboxykinase (PEPCK) in liver. However, PEPCK activity in the liver of sheep subjected to these treatments is not increased (Filsell et al., 1969), and most data from dairy cows indicate that underfeeding or starvation tends to decrease PEPCK activity (Butler and Elliot, 1970).

There are only limited data on glucose homeostasis in carnivores. Nelson et al. (1942) reported that fasting of the carnivorous horned owl up to 7 days produced no change in glucose concentration of the blood. Similarly, Migliorini et al. (1973) reported only comparatively small changes in glucose concentration of blood of another carnivorous bird (black vulture) after three days starvation. Kettelhut et al. (1978) fasted adult cats which had previously been fed either a high-protein low-carbohydrate diet or low-protein high-carbohydrate diet. For the former group of cats, the concentration of glucose ir blood was maintained over a 72 hrs period. For cats previously fed the high-carbohydrate diet glucose concentration fell over the first 24 hours and was then maintained.

Rogers et al. (1977) reported that fasting cats for 5 days caused a significant increase in PEPCK activity of the liver over fed values. However, no significant differences in hepatic PEPCK were found between cats given 17.5 and 70% protein diets. Kettlehut et al. (1978) reported that PEPCK activity of both cytosolic and mitochondrial fractions of cats previously fed high protein diets did not change during fasting. Therefore, PEPCK in the cat fed a high protein diet appears to be non-adaptive as in the ruminant.

Beliveau et al. (1981) examined the comparative rates of gluconeogenesis from and oxidation of ^{14}C labelled glycine, serine and lactate in isolated hepatocytes from rats and cats. Two inhibitors of cytosolic PEPCK (quinolinic acid and mercaptopicolinic acid) depressed gluconeogenesis from serine in rat but not in cat hepatocytes. This suggests that in the cat, serine is converted to glucose by a route not utilizing pyruvate and not involving serine dehydratase. As threonine in excess does not inhibit the conversion of serine to glucose in cat hepatocytes, this indicates different pathways of metabolism for these amino acids. In the rat, both serine and threonine are catabolized via serine dehydratase. These findings taken with earlier reports (Rogers et al., 1977; Rowsell et al., 1979) of low activity of serine dehydratase in the liver of the cat support the alternate pathway of gluconeogenesis via hydroxypyruvate as shown in Fig. 1. Rowsell et al. (1979) reported that the cat has a high hepatic activity of serine-pyruvate aminotransferace and proposed that a high activity of this enzyme is associated in mammals with a flesh-eating dietary habit.

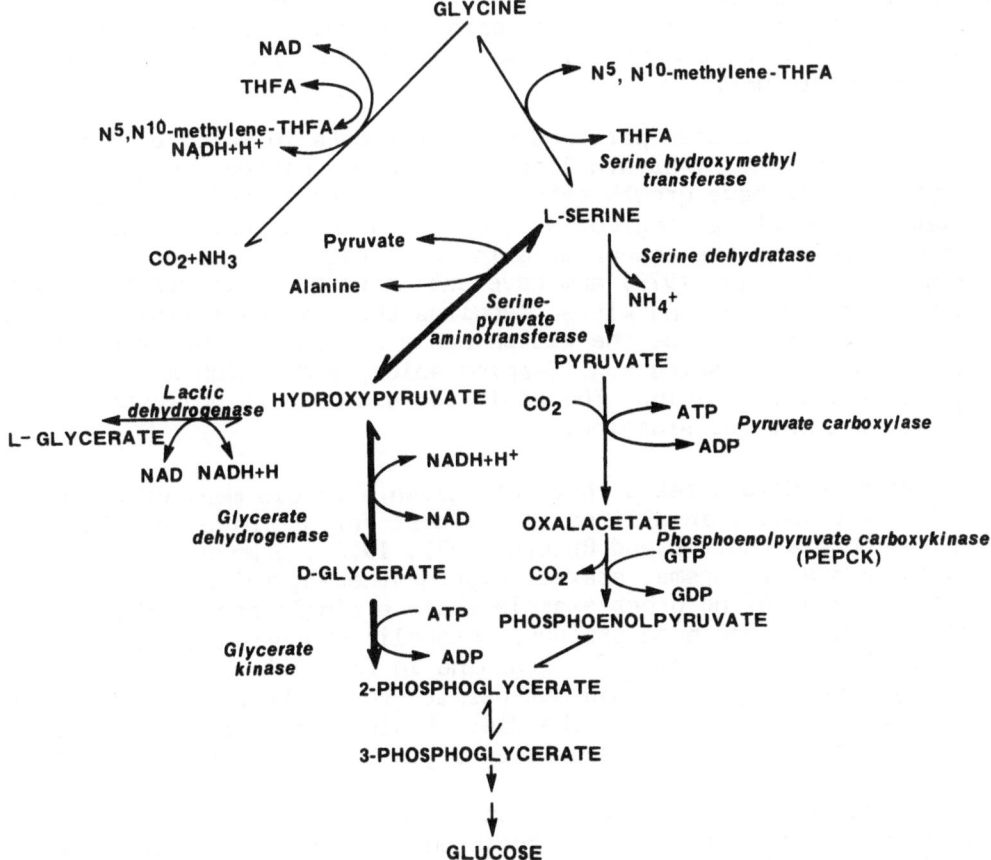

Fig. 1. Gluconeogenesis from serine and glycine showing the alternative pathway (in heavy arrows) not involving serine dehydratase.

AMINO ACID REQUIREMENTS

In the adult ruminant a variable, but generally high proportion of ingested food protein, is degraded in the rumen by the micro-flora to volatile fatty acids and ammonia. Synthesis of microbial protein from carbon skeletons, amino acids and ammonia is an energy linked and dependent process. Microbial protein together with food protein which escapes degradation in the rumen is digested in the intestines of the host. Because of the extensive degradation of food proteins, the profile of amino acids absorbed by the adult ruminant is relatively constant and independent of the amino acid composition of the food ingested. Pregastric fermentation buffers the ruminant against extreme variation in profile of amino acids absorbed, but it places a constraint on the amount absorbed because of the dependency of microbial growth on energy made available during fermentation. The ability of the microflora to utilize ammonia/N for microbial protein synthesis provides the mechanism for utilization of the urea/N entering the rumen via the saliva. The ability to recycle nitrogen may have survival attributes under extreme conditions.

We have developed purified diets for cats in which all protein has been replaced by crystalline amino acids. Kittens when accustomed to these diets have growth rates comparable to those given diets for-mulated from natural ingredients. By using these purified diets we recently defined the essential amino acid requirements of the kitten (Rogers and Morris, 1979) and have estimates of their quantitative re-quirements. The growing kitten requires the same essential amino acids as the growing rat; but the kitten has a somewhat higher requirement for certain amino acids e.g. S-amino acids, and a high and obligatory requirement for arginine. The adult cat also has an obligatory diet-ary requirement for arginine.

We have demonstrated that cats given a single meal of an amino acid diet complete in all respects except for arginine, develop hyperammonemia (Morris and Rogers, 1977, 1978). Symptoms included emesis, muscular spasms, ataxia, hyperaesthesia, tetanic spasms and death. We know of no other example where a single meal lacking a nutrient will cause such profound metabolic changes and even death. Substitution of ornithine for arginine in the diet protected cats from hyperammonemia, but did not permit growth. When citrulline was substituted for arginine in the diet it also prevented hyperammonemia and gave growth rates comparable to arginine (Morris et al., 1979).

These results demonstrate that the cat is unable to synthesize ornithine to maintain the urea cycle when given an amino acid diet devoid of arginine. As adult mammals of most species can synthesize adequate arginine for maintenance, or even positive nitrogen balance, it has generally been assumed that arginine is not a dietary require-ment of adult mammals. The cat (and possibly other carnivores) are in a vulnerable position in that they require arginine not only for

maximal growth, like many other young mammals, but also for maintenance and survival of the adult. The metabolic basis for the
extreme sensitivity of the cat to arginine deprivation has not been
fully elucidated, but the critical steps have been defined.

In mammals, the liver is the primary site of ureagenesis but
does not appear to be involved in production of arginine for other
tissues. The high activity of arginase in the liver precludes release of arginine to the circulation. Rogers et al. (1972) and
Featherstone et al. (1973) suggested that circulating arginine was
synthesized in the kidney from citrulline. Subsequently Windmueller
and Spaeth (1974, 1975) have shown that citrulline was produced in
the small intestine of the rat from glutamine and that 20-30% of the
total plasma glutamine is extracted with each blood pass through the
vasculature (Windmueller, 1980). These authors (Windmueller and
Spaeth, 1981) were unable to demonstrate in the rat any significant
extraintestinal synthesis of citrulline, and showed that under physiological conditions, the liver releases no citrulline. Furthermore,
the majority of the intestinally derived citrulline passes through
the liver without appreciable uptake. But renal uptake of citrulline
was about 83% the rate of intestinal release and the kidneys
reelased arginine equivalent to about 75% of the citrulline taken
up.

Since ornithine prevents hyperammonemia in the cat following in-
gestion of an arginine-free amino acid diet, it appears that the
cat has a reduced capacity to synthesize this compound. Figure 2
shows that γ-glutamyl semialdehyde occupies a key position in the
pathway of both ornithine and proline synthesis. The activities of
the enzyme catalyzing the conversion of γ-glutamyl semialdehyde to
ornithine (ornithine aminotransferase, OAT) in tissues of the rat
and cat were compared by Costello et al. (1981). Activity of OAT
in the liver of the cat is about half that of the rat, similar in
the kidneys and much lower in the intestine of the cat than the rat.
These data indicate that while hepatic synthesis of ornithine from
γ-glutamyl semialdehyde could be comparable in both species, intestinal synthesis would be depressed in the cat. Herzfeld and Raper
(1976) suggested that OAT isolated from the liver of the rat was
less active in the ornithine synthesis than OAT from intestine.
Therefore, OAT in the liver of the cat may function only for degradation of ornithine.

Ross et al. (1978) suggested that in the rat intestine the conversion of glutamate to γ-glutamyl semialdehyde is a two step process:
first involving phosphorylation of glutamate to γ-glutamyl phosphate
by ATP, then reduction of glutamyl phosphate by NADPH to γ-glutamyl
semialdehyde. Subsequently, from the same laboratory, Henslee and
Jones (1981) reported that the reductase of γ-glutamyl phosphate was
located in the mitochondrial fraction of the intestinal mucosa of
the rat. Also, they reported that the mitochondria from rat liver
did not possess reductase activity, despite a high OAT activity.

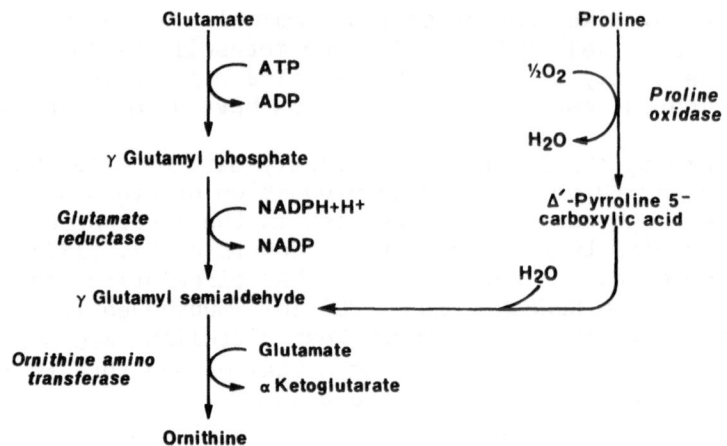

Fig. 2. Ornithine synthesis from glutamate and proline.

This further supports a minor role for the liver of the rat in ornithine synthesis from glutamate. A preliminary report by Costello et al. (1981) suggested that mucosal mitochondria from the cat are deficient in glutamate reductase. If this report of low activities of glutamate, reductase and OAT are confirmed, it will provide an explanation of the extreme sensitivity of the cat to hyperammonemia when given an arginine-free amino acid diet.

REQUIREMENTS FOR VITAMINS, TAURINE AND ESSENTIAL FATTY ACIDS

Adult ruminants are independent of exogenous sources of vitamins, with the exception of vitamins A and E, due to synthesis by the gut microflora. The significance of vitamin synthesis in evolution of pregastric digestion does not appear to have been addressed. However, adoption of coprophagy as a regular practice would achieve similar ends (Hornicke and Bjornhag, 1980).

The vitamin and taurine requirements of the cat provide a marked contrast to those of ruminants and most simple stomached animals.

Vitamin A. Carotene synthesized by plants is the precursor of all naturally occurring vitamin A in animals. In most animals, β-carotene is converted to vitamin A primarily in the intestine. The β-carotene molecule is cleared by a dioxygenase and converted to retinal. Retinal is reduced by another enzyme which requires NADH or NADPH to give vitamin A alcohol (retinol) which is then esterified and absorbed mainly with the chylomicra. Cats are unable to utilize β-carotene as a precursor for vitamin A due to a deficiency of the dioxygenase enzyme. This enzyme deficiency is of minimal consequence to the cat as long as it maintains a carnivorous diet.

Niacin. Most animals obtain their niacin requirements both from preformed dietary nicotinic acid and amides and from the catabolism of tryptophan. For most species the efficiency of conversion of tryptophan to nicotinic acid is about 60:1. Therefore the majority of the tryptophan is metabolized to end products other than nicotinic acid of which those of the glutaryl CoA pathway are most significant. The cat is entirely dependent on preformed niacin in the diet as it is unable to utilize tryptophan for niacin synthesis (DaSilva et al., 1952)

Ikeda et al. (1965) showed that while the cat has a full complement of enzymes of the tryptophan-nicotinic acid pathway, the first unique enzyme of the alternate glutaryl pathway, picolinic carboxylase,has 30-60 greater activity in the liver of the cat than the rat (Suhadolnik et al., 1957). The inability of the cat to utilize the tryptophan-niacin pathway is of minimal consequence as a diet of animal tissue is well supplied with niacin. In contrast, the capacity to rapidly catabolize a tryptophan load assists in removal of tryptophan metabolites; many of which are physiological active compounds.

Taurine. While taurine is not classified as a vitamin, it has been shown to be an essential nutrient for the cat. Chemically it is a β-sulfonic amino acid. The magnitude of the dietary taurine requirement for the cat is near the vitamin range (less than 0.05% of the diet).

A dietary deficiency of taurine in the cat was shown by Hayes et al. (1975) to be associated with the development of central retinal degeneration. The retina contains a high concentration of taurine which is required for maintenance of its integrity. Concentrations of taurine in the tissues of the cat are determined by the balance between dietary intake, synthesis, excretion and changes in size of pools. The cat is capable of synthesizing limited amounts of taurine from the sulphur amino acids, methionine and cystine (Knopf et al., 1978). The metabolic basis for this limited synthesis has not been fully elucidated but is presumed to be due to the low, non-adaptive activity of cysteinesulfinic acid decarboxylase.

For many placental mammals, the bile salts contain a mixture of the taurine and glycine conjugates of cholic acid. When confronted with a deficiency of taurine, glycocholic acid becomes the major conjugate. Rabin et al. (1976) showed that the cat is unable to alternate between the glycine and taurine conjugation. The proportion of glycocholic acid does not increase when there is a tissue deficit of taurine; that is, the cat conjugates cholic acid only with taurine. Therefore, the cat has a continual need for taurine for bile salt formation and is unable to conserve enough taurine to prevent a deficiency state.

Recently we have shown that the concentration of sulfur amino acids in the diet markedly affects the concentration of taurine in plasma and the subsequent incidence of central degeneration. Dietary and other factors affecting the efficiency of taurine recovery in the enterohepatic circulation also probably influence taurine status.

Although taurine does not form a peptide bond, it is associated with all animal tissues, being particularly high in the liver. A carnivore consuming a diet of animal tissue would be absorbing not only preformed taurine but also a high intake of S-amino acids, the precursor for taurine synthesis. Therefore, a carnivorous diet would never be associated with a functional taurine deficiency.

Essential fatty acids. In ruminants, dietary lipids are extensively modified in the rumen before absorption in the intestine. Galactolipids and triglycerides are hydrolyzed; the glycerol and galactose moieties being fermented to volatile fatty acids. Liberated unsaturated fatty acids from lipids are extensively hydrogenated by certain bacteria, and double bonds are altered from the *cis* to the *trans* position. Presumably sufficient unsaturated fatty acids escape hydrogenation in the rumen to provide the requirements of the host.

Linoleic and linolenic acids are usually regarded as the essential fatty acids being used directly in phospholipids and also providing precursors for chain elongation and desaturation reactions which result in the production of extremely active physiological compounds such as prostaglandins, etc. Recent studies indicate that some carnivorous animals have limited ability to undertake desaturation of fatty acids. Rivers and associates (Rivers et al. 1975, 1976a; Hassam et al., 1977; Frankel and Rivers, 1978) suggested that the cat lacked both the Δ6 and Δ8 desaturases and hence had a requirement for arachedonic acid in the diet. They also suggested (Rivers et al., 1976b) that the lion may also lack the Δ6 and Δ8 desaturases.

This metabolic anomaly would have no deleterious effect on the fitness of carnivores as long as they maintained a carnivorous diet.

CONCLUSIONS

The examples discussed in this paper indicate that the metabolic machinery of the ruminant and true carnivore have adapted to the peculiar end products of their digestive systems. By deletion of enzymes rendered redundant by their strict adherence to an animal tissue diet, carnivores have eliminated the need to synthesize proteins (enzymes) which appear to be of no advantage to the animal.

In ruminants, microbial fermentation of food results in little glucose being absorbed from the gut. Ruminant metabolism has been directed towards the use of acetate rather than glucose for fatty acid synthesis and a constant high rate of gluconeogenesis.

ACKNOWLEDGMENT

We wish to acknowledge financial support of the Carnation Company, Los Angeles, California.

REFERENCES

Ballard, F. J., 1965, Glucose utilization in mammalian liver, Comp. Biochem. Physiol., 14:437-443.

Bauman, D. E., and Davis, C. L., 1975, Regulation of lipid metabolism, in: "Digestion and Metabolism in the Ruminant," I.W. McDonald, and A.C.I. Warner, eds., Univ. New Engl. Publ. Unit., Armidale, Australia.

Bedford, P. G. C., and Clark , E. G. C., 1972, Experimental benzoic acid poisoning in the cat, Vet. Rec., 90:53-58.

Beliveau, G. P., Morris, J. G., Rogers, Q. R., and Freedland, R. A., 1981, Metabolism of serine, threonine and glycine in isolated cat hepatocytes, Fed. Proc., 40:807, Abstr. No. 3504.

Butler, T. M., and Elliot, J. M., 1970, Effect of diet and glucocorticoid administration on liver phosphoenolpyruvate carboxykinase activity in the dairy cow, J. Dairy Sci., 53:1727-1733.

Colbert, E. H., 1980, "Evolution of the vertebrates," 3rd edn., John Wiley & Sons, New York.

Costello, M. J., Morris, J. G., and Rogers, Q. R., 1981, The role of intestinal mucosa in endogenous arginine biosynthesis in ureotelic mammals, XIIth Int. Cong. Nutr., Aug. 16-21, 96 Abstr 538, San Diego, Ca.

DaSilva, A. C., Fried,R., and DeAngelis, R. C., 1952, The domestic cat as a laboratory animal for experimental nutrition studies. III. Niacin requirement and tryptophan metabolism, J. Nutr., 46:399-409.

Dunlop, R. H., and Hammond, P. B., 1965, D-lactic acidosis of ruminants, Ann. NY Acad. Sci., 119:1109-1132.

Featherston, W. R., Rogers, Q. R., and Freedland, R. A., 1973, Relative importance of kidney and liver in synthesis of arginine by the rat, Am. J. Physiol., 224:127-129.

Filsell, O. H., Jarrett, I. G., Taylor, R. H., and Keeth, D. B.,

1969, Effects of fasting, diabetes and glucocorticoids on gluconeogenic enzymes in the sheep, Biochim. Biophys. Acta, 184:54-63.

Frankel, T. L., and Rivers, J. P. W., 1978, The nutritional and metabolic impact of γ-linolenic acid (18:3ω6) on cats deprived of animal lipid, Brit. J. Nutr., 39:227-231.

Freeland, W. J., and Janzen, D. H., 1974, Strategies in herbivory by mammals: the role of plant secondary compounds, Am. Nat., 108:269-289.

Geist, V., 1972, An ecological and behavioural explanation of mammalian characteristics and their implication to therapsid evolution, Z. Saugetierkunde, 37:1-5.

Giesecke, D., and Stangassinger, M., 1980, Lactic acid metabolism, in:"Digestive physiology and metabolism in ruminants,"Y. Rucker-busch and P. Thivend, eds., MTP Press Ltd., Lancaster, England.

Harper, A. E., 1965, Effect of variation in protein intake on en-zymes of amino acid metabolism, Can. J. Biochem., 43:1589-1603.

Hassam, A. G., Rivers, J. P. W., and Crawford, M. A., 1977, The failure of the cat to desaturate linoleic acid; its nutritional implications, Nutr. Metab., 21:321-328.

Hayes, K. G., Carey, R. E., and Schmidt, S. Y., 1975, Retinal de-generation associated with taurine deficiency in the cat, Science, 188:949-950.

Henslee, J. G., and Jones, M. E., 1981, Ornithine synthesis from glutamate in mitochondria from rat small intestinal mucosa, Fed. Proc., 40:1683, Abstr. No. 829.

Herzfeld, A., and Raper, S. M., 1976, Enzymes of ornithine metabolism in adult and developing rat intestine, Biochim. Biophys. Acta, 428:600-610.

Hirom, P. C., Idle, J. R., and Millburn, P., 1977, Comparative aspects of the biosynthesis and excretion of xenobiotic conju-gates by non-primate mammals, in:"Drug metabolism--from microbe to man," D.V. Parke,and R.L. Smith, eds., Taylor and Francis Ltd., London.

Hornicke, H., and Bjornhag, G., 1980, Coprophagy and related strate-gies for digesta utilization, in:"Digestive physiology and metabolism in ruminants," Y. Ruckerbusch and P. Thivend, eds., MTP Press Ltd., Lancaster, England.

Ikeda, M., Tsuji, H., Nakamura, S., Ichiyama, A., Nishizuka, Y., and Hayaishi, O., 1965, Studies on the biosynthesis of nicotinamide adenine dinucleotide. II. A role of picolinic carboxylase in the biosynthesis of nicotinamide adenine dinucleotide from tryptophan in mammals, J. Biol. Chem., 240:1395-1401.

James, L. F., Allison, M. J., and Littledike, E. T., 1975, Produc-tion and modification of toxic sybstances in the rumen, in: "Digestion and metabolism in the ruminant," I.W. McDonald and A.C.I. Warner, eds., Univ. New Eng. Publ. Unit., Armidale, Australia.

Kaneko, J. J., 1980, "Clinical biochemistry of domestic animals," 3rd edn., Academic Press, New York.

Kettelhut, I. C., Foss, M. C., and Migliorini, R. H., 1978, Glucose
    homeostasis in a carnivorous animal (cat) and in rats fed a
    high-protein diet, Am. J. Physiol., 239. (Regulatory Integrative
    Comp. Physiol. 3) R115-R121.
Knopf, K., Sturman, J. A., Armstrong, M., and Hayes, K. C., 1978,
    Taurine: an essential nutrient for the cat, J. Nutr., 108:
    773-778.
Knox, W. E., Greengard, O., 1965, The regulation of some enzymes of
    nitrogen metabolism--an introduction to enzyme physiology,
    Adv. Enzyme Regulation, 3:247-313.
Liang, Y. T., Morill, J. L., and Noordsy, J. L., 1967, Absorption
    and utilization of volatile fatty acids by young calves,
    J. Dairy Sci., 50:1153.
Lindsay, D. B., 1978, Gluconeogenesis in ruminants, Biochem. Soc.
    Trans., 6:1152-1156.
Lindsay, D. B., 1982, Relationships between amino acid catabolism
    and protein anabolism in the ruminant, Fed. Proc., 41:2550-2554.
McCandless, E. L., and Dye, J. A., 1950, Physiological changes in
    intermediary metabolism of various species of ruminants inciden-
    tal to functional development of the rumen, Am. J. Physiol.,
    162:434-446.
Migliorini, R. H., Linder, C., Moura, J. L., and Veiga, J. A. S.,
    1973, Gluconeogenesis in a carnivorous bird (black vulture),
    Am. J. Physiol., 225:1389-1392.
Morris, J. G., and Rogers, Q. R., 1977, Ammonia intoxication in the
    near-adult cat as a result of a dietary deficiency of arginine,
    Science, 199:431-432.
Morris, J. G., and Rogers, Q. R., 1978, Arginine: An essential amino
    acid for the cat, J. Nutr., 108:1944-1953.
Morris, J. G., Rogers, Q. R., Winterrowd, D. L., and Kamikawa, E. M.,
    1979, The utilization of ornithine and citrulline by the grow-
    ing kitten, J. Nutr., 109:724-729.
Nelson, N., Elgart, S., and Mirsky, I. A., 1942, Pancreatic diabetes
    in the owl, Endocrinology, 31:119-123.
Rabin, B., Nicolosi, R. J., and Hayes, K. C., 1976, Dietary influence
    on bile acid conjugation in the cat, J. Nutr., 106:1241-1246.
Rivers, J. P. W., Sinclair, A. J., and Crawford, M. A., 1975, In-
    ability of the cat to desaturate essential fatty acids, Nature,
    258:171-173.
Rivers, J. P. W., Hassam, A. G., and Alderson, C., 1976a, The absence
    of Δ6-desaturase activity in the cat, Proc. Nutr. Soc., 35:67A.
Rivers, J. P. W., Hassam, A. G., Crawford, M. A., and Brambell, M. R.,
    1976b, The inability of the lion Panthera léo L. to desaturate
    linoleic acid, FEBS Letters, 67:269-270.
Rogers, Q. R., Freedland, R. A., and Symmons, R. A., 1972, In vivo
    synthesis and utilization of arginine in the rat, Amer. J.
    Physiol., 223:236-240.
Rogers, Q. R., Morris, J. G., Freedland, R. A., 1977, Lack of hepatic
    enzymatic adaptation to low and high levels of dietary protein
    in the adult cat, Enzyme, 22:348-356.

Rogers, Q. R., and Morris, J. G., 1980, Why does the cat require a high protein diet? in:"Nutrition of the dog and cat," R.S. Anderson, ed., Pergamon Press, Oxford.

Romer, A. S., 1966,"Vertebrate paleontology," 3rd edn., Univ. Chicago Press, Chicago.

Ross, G., Dunn, D., and Jones, M. E., 1978, Ornithine synthesis from glutamate in rat intestinal mucosa homogenates. Evidence for the reduction of glutamate to γ-glutamyl semialdehyde, Biochem. Biophys. Res. Commun., 85:140-147.

Rowsell, E. V., Carnie, J. A., Wahbi, S. D., Al-Tai, A. H., and Rowsell, K. V., 1979, L-serine dehydratase and L-serine-py-ruvate aminotransferase activities in different animal species, Comp. Biochem. Physiol., 63B:543-555.

Schimke, R. T., 1962, Adaptive characteristics of urea cycle enzymes in the rat, J. Biol. Chem., 237:459-468.

Scharrer, E., Riegger, A., Amann, B., Weidl, G., and Seeliger, U., 1976, Developmental changes of sugar transport in ovine small intestine, Pfluger's Arch., 366:147-151.

Suhadolnik, R. J., Stevens, C. O., Decher, R. H., Henderson, L. M., and Hankes, L. V., 1957, Species variation in the metabolism of 3-hydroxyanthranilate to pyridinecarboxylic acids, J. Biol. Chem., 228:973-982.

Thivend, P., Toullec, R., and Guilloteau, R., 1980, Digestive adaptation in the preruminant, in:"Digestive physiology and metabolism in ruminants," Y. Ruckerbusch,and P. Thivend, eds., MTP Press Ltd., Lancaster, England.

Toofanian, F., Hill, F. W. G., and Kidder, D. E., 1973, The mucosal disaccharidases in the small intestine of the calf, Ann. Rech. Vet., 4:57-69.

Vandersall, J. H., Conrad, H. R., and Hibbs, J. W., 1957, Relation of life span of red blood cells tagged with chromium-51 to corpuscle glucose in youngcalves, J. Anim. Sci., 16:1111.

VanSoest, P. J., 1982, "Nutritional ecology of the ruminant," O & B Books Inc., Oregon.

Windmueller, H. G., and Spaeth, A. E., 1974, Uptake and metabolism of plasma glutamine by the small intestine, J. Biol. Chem., 249:5070-5079.

Windmueller, H. G., and Spaeth, A. E., 1975, Intestinal metabolism of glutamine and glutamate from the lumen as compared to glutamine from blood, Arch. Biochem. Biophys., 171:662-672.

Windmueller, H. G., and Spaeth, A. E., 1981, Source and fate of circulating citrulline, Am. J. Physiol., 241 (Endocrinol. Metab. 5): E473-E-480.

Windmueller, H. G., 1980, Enterohepatic aspects of glutamine metabolism, in:"Glutamine metabolism enzymology and regulation," J. Mora and R. Palacios, eds., Academic Press Inc., New York.

SUPEROXIDE DISMUTASE ACTIVITY IN THE RAT EXPOSED TO EXTREME

ENVIRONMENTAL CONDITIONS

V. M. Petrović, M. Spasić, B. Milić, Z. Saičić, and
R. Radojičić

Institute of Biological Research and Institute of
Physiology and Biochemistry, Faculty of Sciences,
University of Belgrade, Belgrade, Yugoslavia

INTRODUCTION

In spite of the fact that most of the oxygen consumed by respiring cells is utilized by cytochrome oxidase which reduces oxygen to water without releasing either O_2^- or H_2O_2 (Antonini et al.,
1970), these radicals are made in respiring cells in a considerable
amount. It has been shown that autooxidation of hydroquinones,
leukoflavins, catecholamines, thiols and tetrahydropterins generate
O_2^-. The superoxide radicals have been demonstrated as an intermediate in the mechanisms of action of galactose oxidase, indoleamine
dioxygenase and 2-nitropropane dioxygenase (Fridovich, 1978). Frag-
ments of subcellular organells such as mitochondria and chloro-
plasts also produce O_2^-. These radicals have been shown to induce
lipid peroxidation, damage membranes and kill cells. A more potent
oxidant H_2O_2 is produced in the metabolic pathway to the membrane
systems whose structure is currently viewed as a mosaic array of
protein particles floating in a double-layered sheet of lipid
molecules. It has been shown that this structure responses to
ionizing radiation by governing both the protein and lipid moities.
The whole body exposure of mice to sub-lethal doses of X-rays is
followed by destruction of the white blood cells and by a decline
in their number. Superoxide radicals are produced during irradiation
of oxygenated tissues and these radicals may be scavenged by superoxide dismutase (Mitsuyuki et al., 1981; Bartosz et al., 1979;
Petkau et al., 1976, 1978). Ionizing radiation enhances the development of carcinogenesis in various tissues, which is localized mostly
in lungs, breasts and ovaries (Kalisnik et al., 1981).

Superoxide dismutase is a histoprotective enzyme which catalyzes the dismutation or disproportiation of superoxide free radical anions, yielding hydrogen peroxide and oxygen. The following pathway of oxygen reduction and enzymatic defence against the intermediates has been presented by Fridovich (1978).

$$O_2 \xrightarrow{\;e^-\;} O_2^- \xrightarrow{\;e^- + 2H^+\;} H_2O_2 \xrightarrow[H_2O]{\;e^- + H^+\;} OH \xrightarrow{\;e^- + H^+\;} H_2O$$

$$O_2^- + O_2^- + 2H^+ \longrightarrow H_2O_2 + O_2 \qquad \text{Superoxide dismutase}$$

$$H_2O_2 + H_2O_2 \longrightarrow 2H_2O + O_2 \qquad \text{Catalases}$$

$$H_2O_2 + RH_2 \longrightarrow 2H_2O + R \qquad \text{Peroxidases}$$

Superoxide radicals are eliminated by superoxide dismutases, which catalyze their conversion to hydrogen peroxide, plus oxygen. Hydrogen peroxide is removed by catalases, which convert it to water plus oxygen and by peroxidases, which reduce it to water.

Many examples could demonstrate the serious damage which is a result of biological systems being exposed to a source of superoxide production. The phenomenon can be prevented by addition of superoxide dismutase, but it can be prevented by the addition of catalase, too. Some results indicate that a complete protection may be obtained with the addition of either superoxide dismutase or catalase (McCord, 1979). We have recently found that the activity of this protective enzyme (SOD) may be stimulated by noradrenaline or thyroid hormones (Petrović et al., 1980, 1981, 1982). In the present experiments we studied the effect of some extreme external conditions on superoxide dismutase activity in various tissues of rat.

MATERIALS AND METHODS

Male Mill Hill hooded rats were placed in individual cages at mass of 170-200 g and maintained either at 22-24°C or 5±1°C with 12 hrs lighting schedule for up to 5 months, as described in each experiment. They had free access to food and water.

In the second experiment rats were irradiated with a Roenthgen apparatus - Phillips RT 250 operated at 200 KW, 20 mA, 0.5 mm Cu. The concentration of free fatty acids in the blood and activity of superoxide dismutase in the tissues were examined in the irradiated animals.

Animals were exposed to the following doses of X-rays: 0.14 C/kg and sacrificed 15 or 60 min after irradiation; 0.19 C/kg and sacrificed 15, 60 or 180 min after irradiation. The dose-rate was 1.9604 x 10^{-2} C/kg/min, the target distance was 50 cm.

In those experiments in which we investigated the effect of X-irradiation on SOD activity, rats were exposed to a dose of 0.20 C/kg, and dose-rate 2.0124 x 10^{-2} C/kg/min, the target distance was 51.5 cm.

Concentration of FFA in the serum was determined by the method of Ducombe (1964).

Rats were killed by decapitation between 8-10 a.m. Interscapular brown adipose tissue (IBAT) was removed, placed in ice cold 0.9% NaCl solution and carefully trimmed the adhering skeletal muscle and white adipose tissue. Once washed, tissue were blotted with a filter paper and weighed. Liver, and brain were removed within 3 minutes, the liver being perfused with ice-cold 0.9 NaCl solution prior to the removal. After being removed the tissues were minced and then dispersed with a loosely-fitted Potter-Elvehjem pestle in nine volumes of the buffer, containing 0.05 M KH_2PO_4 and 10^{-4} M EDTA, pH 7.8. All the operations were performed at $4^{o}C$. Homogenate was centrifuged for 15 min at 10,000 r.p.m. in Sorvall centrifuge. Supernatant was centrifuged for 90 min at 37,000 r.p.m in Beckman centrifuge and then used for the determination of SOD activity, as described by Misra and Fridovich (1972). Protein was determined according to the method of Lowry et al. (1951). Spectrophotometric assays were performed in a Gilford model 250. All reagents used were proanalysis.

RESULTS

Effect of acute and chronic exposure to cold

As it is shown in Fig. 1 in animals previously adapted to 22-$24^{o}C$ and transferred to cold ($5\pm1^{o}C$), for 6 hours, cytosol SOD activity in the IBAT was found to be lower than in controls kept at $20-24^{o}C$ ($p < 0.01$). After the exposure to cold for 21 days, the activity of this enzyme was found to be at the level of controls. However, in animals adapted to cold (5 months of exposure) super-oxide dismutase activity was significantly increased ($p < 0.01$).

The mass of IBAT (Table 1) was significantly increased in animals exposed to cold for 21 days as well as in animals adapted to cold (5 months of exposure), but after 6 hours of exposure, the mass was at the control level. The amount of isolated cytosol protein (expressed per g wet mass tissue) was higher in animals exposed to cold in respect to the control, independent of the duration of exposure.

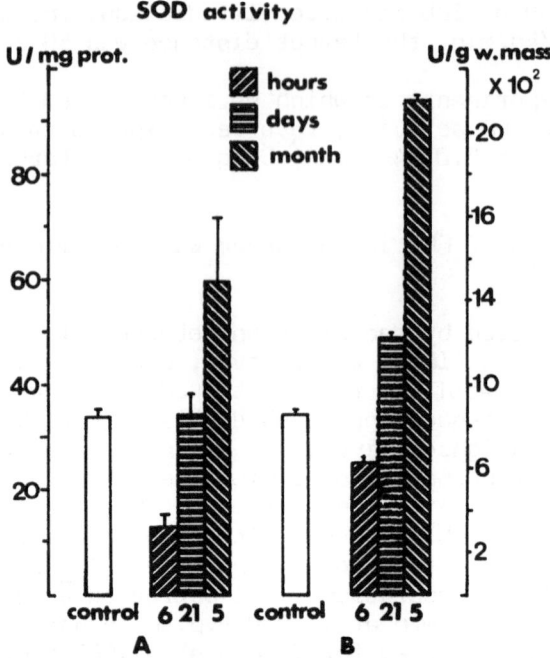

Fig. 1. Superoxide dismutase activity in brown adipose tissue of
        rats exposed or adapted to cold expressed in units per mg
        of protein (A) or per g wet tissue (B). Means ± SE means
        of 6-8 animals. Control: 6 hours or 5 months of exposure
        p < 0.01.

Table 1. IBAT mass and protein content in the rats exposed or adapt-
         ed to cold.

	RATS EXPOSED OR ADAPTED TO COLD			
	Control	Hours 6	Days 21	Months 5
n	31	6	8	8
IBAT mass (g)	0.25	0.26	0.53	0.91
$\frac{mg\ protein}{g\ wet\ mass}$	25±1	49±4	36±3	34±4

If expressed in units per g wet mass tissue (Fig. 1B), it may be observed that the activity of this enzyme was lower after 6 hours of exposure but, significantly higher in animals exposed to cold either for 21 days or 5 months.

In the liver of rats exposed to cold for 6 hours SOD activity was increased ($p < 0.01$). Significantly higher activity was also found in the cold adapted animals, but it was only slightly higher after 21 days of exposure (Fig. 2).

In the brain, no significant changes in SOD activity in animals exposed or adapted to cold were found (Fig. 3).

Effect of X-irradiation

We have determined the concentration of free fatty acids (FFA) in blood serum of animals exposed to different X-ray doses (1.4, 1.9 and 2.7 C/kg). As shown in Fig. 4 the doses of 1.9 or 2.7 C/kg produced a significant increase in FFA concentration. This effect was already registered 15 minutes after the X-irradiation with maximum effect expressed after 60 minutes.

So we decided to study SOD activity in the liver, brain and IBAT of animals killed 60 minutes after the whole body exposure to X-rays of 1.9 C/kg. As shown in Fig. 5 a significant increase in

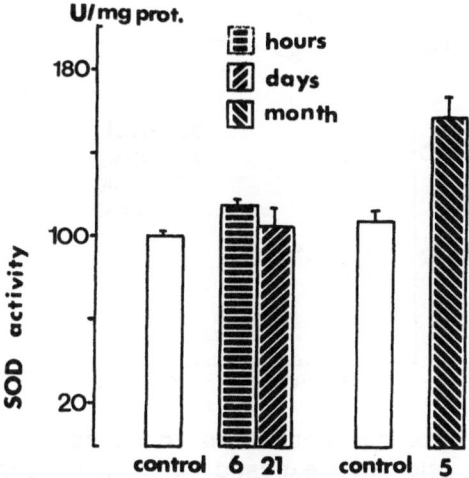

Fig. 2. Superoxide dismutase activity in the liver of rats exposed or adapted to cold (U/mg protein). Control: 6 hours or 5 months of exposure $p < 0.01$.

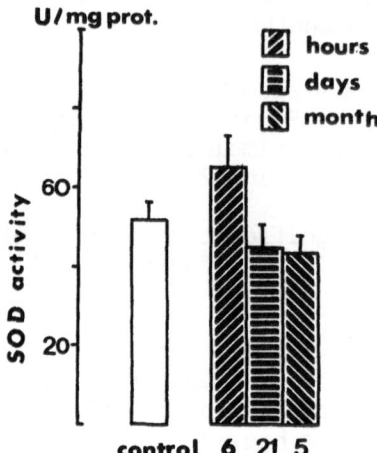

Fig. 3. Superoxide dismutase activity in the brain of rats exposed or adapted to cold. Differences between control and cold exposed animals p < 0,05.

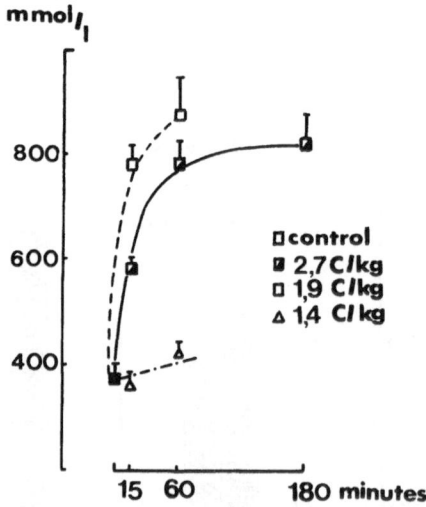

Fig. 4. Free fatty acid concentration in blood serum (m mol/l) in the control and rats exposed to X-rays. Control: irradiated 2,7 C/kg and 1,9 C/kg p < 0,01.

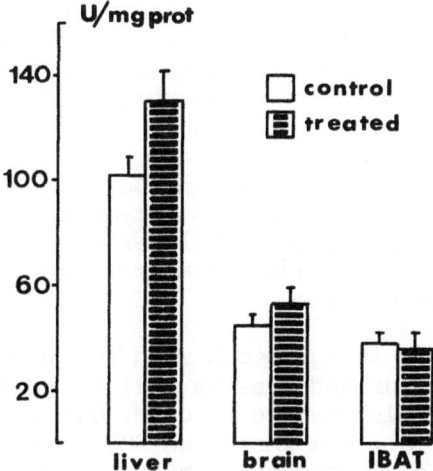

Fig. 5. Superoxide dismutase activity in the liver, brain and IBAT
of control and irradiated rats. For the liver the difference between control and irratiated animals is significant
p < 0,01.

SOD activity was found only in the liver, but in the brain and IBAT
of X-irradiated rats no changes were observed.

DISCUSSION

     Exposure of mammals to temperatures below their thermoneutral
zones led to increased generation of heat by shivering and/or non-
shivering pathways. It was demonstrated that shivering was manifested only by muscle while nonshivering thermogenesis reflected changes
in the metabolic activities of several organ systems, especially
interscapular brown adipose tissue, muscle and liver (Jansky, 1973).
Changes in IBAT metabolism during exposure or adaptation to cold
were intensively examined in the past few years (Depocas et al., 1980;
Himms-Hagen et al., 1980; Horwitz, 1979). In the first 24 hours of
exposure to cold, changes in IBAT may be described as acute cold
stress.

     Our present data confirm some previous findings which concern
changes in the mass and the protein content in IBAT of animals
exposed to cold (Himms-Hagen, 1972; Horwitz, 1974; Jansky, 1973,
1976). The color of this tissue was clear brown in cold exposed
animals and pale brown in heat exposed ones, which may be due to

the changes in such colored substances as flavoproteins, electron transport elements and better blood supply. In addition many more mitochondria and higher specific activity of oxidative enzymes than in heat was found in the cold ambient (Chaffee and Roberts, 1971). The experiment on monkeys (*Macaca mulatta*) has shown a significant difference in the mass and protein content of axillary brown adipose tissue between cold and heat exposed animals (Chaffee et al., 1971). The total calculated enzymatic activity in BAT associated with succinate, α-kettoglutarate, β-hydroxybutyrate and α-glycerophosphate of cold adapted monkeys was significantly higher than in heat exposed animals.

Our findings are in agreement with the above mentioned results in respect to the color and mass as well as to the amount of protein. In our animals adapted to cold, but also in the 21 days exposed ones (Fig. 1B), the increased SOD activity may be due to the increased flux of O_2 radicals, but some endogen hormones may be involved too.

It is well known that noradrenaline induces an increase in the metabolic rate of cold adapted animals. In the cold ambient there is an increase in catecholamines secretion which might influence the respiratory rate of IBAT.

A model has been developed which gives considerable insight into how noradrenaline might influence the respiratory rate of this tissue (Horwitz, 1974). Noradrenaline stimulates BAT respiration by causing an increase in the cellular availability of ADP in two ways. It depolarizes the BAT cell membranes and thus changes its sodium permeability and activates adenyl cyclase so that there is an increase in CAMP. Both these effects of noradrenaline cause the activation of Na^+/K^+ ATP-ase (Horwitz, 1974). The resulting increase in ADP formation, coupled with a noradrenaline induced lipolysis in BAT (and also in white adipose tissue), which increases intracellular availability as well as the circulating levels of fatty acids, suggests a mechanism whereby the high BAT respiratory rate induced by cold stress of noradrenaline injection, may be sustained. Adding to this cellular metabolic release is the fact that fatty acids can directly loose the oxidative phosphorylation of BAT mitochondria (Himms-Hagen, 1972).

In our present experiment the increase in SOD activity may be due at least to two factors, one of them being the action of noradrenaline on respiratory systems in IBAT and consequently on the production of O_2 radicals which may induce SOD activity. However, the direct hormonal action should be also taken into consideration. In fact we have shown that noradrenaline may produce a significant increase in SOD activity in IBAT, liver and brain. However, thyroid hormones produced a significant stimulatory effect only in the brain (Petrović et al., 1980, 1981).

In spite of the fact the liver represents only a small percentage of the body mass it has been estimated to account for more
than 20% of the metabolic rate in the rat (Jansky, 1963, 1966; Hoch,
1971). In the cold an increase in the activity of oxidative enzymes
and change in the high energy bound phosphate formed per gram -
atom of oxygen consumed (P:O ratio) occur in the liver mitochondria
(Hoch, 1971).

In the liver microsomes of cold acclimated rodents the dihydro-
nicotinamide adenine dinucleotide phosphate (NADPH) and dihydro-
nicotinamide adenine dinucleotide (NADH) - cytochrome C reductase
are increased (Chaffee, 1976). Cytochrome C and coenzyme Q have
been found to be increased in this condition too (Beyer, 1971;
Depocas, 1966). There is also evidence that the liver utilizes more
glucose in the cold since a metabolite of glucose degradation such
as phospho-enol-pyruvate was found in higher concentration than in
heat (Cassuto et al., 1970). Thus in some ways the liver glycolytic
responses in cold acclimated rodents are opposite to those found in
heat acclimated ones.

Thyroid activity which is known to be increased in the cold is
presumably the cause of the increased activity of liver Na^+/K^+ ATP-
ase. We found that the specific activity of cytosol and mitochondrial
proteins (expressed in Cpm/mg proteins) after exposure to cold for
3, 6 or 12 hours and for 1, 2 or 7 days was higher than in control
adapted to $24^\circ C$. In animals exposed to cold for 14 or 21 days, this
activity was at the level of controls. After a prolonged exposure
to cold (21 days) the liver mass was increased and if expressed
per total liver mass, the radioactivity of cytosol and mitochondrial
proteins was higher in cold, indicating an increased capacity of
the liver for protein synthesis (Petrović at al., 1979). Since the
rate of protein synthesis is dependent on thyroid activity, the
fact that liver ribosomes from cold acclimated animals incorporate
amino acids into protein at a higher rate than do ribosomes from
controls, may be attributed to the action of thyroid hormones
(Chaffee, 1968; Hoch, 1971).

Our finding which concern the increased SOD activity in the
liver of rats exposed to cold for 6 hours may be due to the increased
adrenocortical activity found in that conditions. However, the highest activity of this enzyme registered in animals exposed to cold
for 5 months should be attributed to the increased thyroid activity
(Petrović, 1976). SOD activity in the brain was not changed under
influence of cold.

Environmental radiation is known to induce a cellular response
which may result in cancer; the biomembranes are radiation sensitive
structures (Wallach and Weidekamm, 1973; Petkau, 1980). Irradiated
lipids undergo changes in composition with a decrease in normal

unsaturated fatty acids and phospholipids and an increase in peroxidation products (Schwarz et al., 1963; Hammer and Wills, 1979).
These products activate potential chemical carcinogens, such as
N-hydroxy-N-acetyl-2-aminofluorene(via free radical processes)
which react with microsomal membranes as well as with lipids containing carbon-carbon double bonds and produce tumors (Floyd et al.,
1976, 1978; Floyd, 1977). Taking into consideration these facts,
the conversion and elimination of superoxide radicals from the
tissue is of exceptional importance. It was shown that bovine superoxide dismutase protected x-irradiated mice and the protection
occurred in a dose range where hematological damage was an important
contributor to animal lethality (Petkau, 1978).

Our present results which concern the increase in free fatty
acids concentration in the blood of x-irradiated animals are in
agreement with previous findings (Petkau, 1980). As far as SOD
activity is concerned, this enzyme in x-irradiated animals was
changed only in the liver but not in IBAT, contrary to our previous
finding when the effect of noradrenaline was evaluated. The low
doses of this agent produced an increase in SOD activity only in
IBAT but higher doses exerted a significant effect in both, the
liver and IBAT (Petrović et al., 1980, 1981). Taking into consideration all these facts we may suggest that in x-irradiated animals
some other mechanism is involved, different from hormonal, effect-
ing SOD activity only in the liver. For answering this question,
additional experiment should be undertaken.

In conclusion, SOD seems to play an important role as a protec-
tor when animals are exposed to the extreme environmental conditions
such as acute and chronic exposure to cold or to x-irradiation. This
histoprotective enzyme defends cells against the superoxide radicals
and is necessary for survival. However, the precise mechanism by
which the external factors induce the synthesis and activation of
SOD should be further elucidated.

REFERENCES

Antonini, E., Brunori, M., Grenwood, C., and Malmstrom, M., 1970,
     Catalytic Mechanism of Cytochrome Oxidase, Nature, 228:936-937.
Bartosz, G., Leyko, W., and Fried, R., 1979, Is superoxide dismutase
     a physiological radioprotector?, Experientia, 35:1194
Beyer, R. E., 1971, Biochemical aspects of acclimation to a cold
     environment, in: "Biochemical responses to environmental stress,"
     I. A. Bernstein, ed..
Cassuto, Y., Chayoth, R., and Rabi, T., 1970, Thyroid hormone in
     heat acclimated hamsters, Am. J. Physiol., 218:1287-1290.
Chaffee, R. R. J., and Roberts, J. G., 1971, Temperature regulation
     in birds and mammals, Ann. Rev. Physiol., 33:155-202.
Chaffee, R. R. J., 1976, Effect of heat on cellular metabolism:

enzymes and other cellular changes, in:"Progress in animal bio-
    meteorology," E. Tromps, ed., Swets and Zeitlinger Publ. Co.,
    Amsterdam.
Depocas, F., 1966, Concentration and turnover of cytochrome C in
    skeletal muscles of warm and cold-acclimated rats, Can. J.
    Physiol. Pharmacol., 44:875-880.
Depocas, F., Foster, D. O., and Lorraine-Frydman, M., 1980, Nor-
    adrenaline-induced calorigenesis in warm-and in cold-acclimated
    rats: relations between concentration of noradrenaline in
    arterial plasma, blood flow to differently located masses of
    brown adipose tissue and calorigenic response, Can. J. Physiol.
    Pharmacol., 58(8):915-925.
Ducombe, W. G., 1964, The calorimetric microdetermination of non-
    esterified fatty acids in plasma, Clin. Chim. Acta, 9:122-125.
Floyd, R. A., Soong, L. M., Walker, R. N., and Stuart, M., 1976,
    Lipid hydroperoxide activation of N-hydroxy-N-acetylaminofluo-
    rene via a free radical route, Cancer Research, 36:2761-2767.
Floyd, R. A., 1977, Free radicals produced by nitrosofluorene - un-
    saturated lipid reaction, Experientia, 33:197-198.
Floyd, R. A., Soong, L. M., Stuart, M., and Reigh, D. L., 1978, Free
    radicals and carcinogenesis, Arch. Biochem. Biophys., 185:
    450-457.
Fridovich, I., 1978, The biology of oxygen radicals, Science, 201:
    875-880.
Hammer, C. T., and Wills, E. D., 1979, The effect of ionizing
    radiation on the fatty acid composition of natural fats and on
    lipid peroxide formation, Int. J. Radiat. Biol., 35:323-332.
Himms-Hagen, J., 1972, Lipid metabolism during cold exposure and
    during cold acclimation, Lipids, 7:310-323.
Himms-Hagen, J., Dittmar, E., Zaror-Behrens, G., 1980, Polypeptide
    turnover in brown adipose tissue mitochondria during acclimation
    rats to cold, Can. J. Biochem., 58:336-344.
Hoch, E. L., ed., 1971, "Energy transformation in mammals," Sounders
    W.B. Co., Philadelphia, London, Toronto.
Horwitz, B. A., 1974, Physiological and biochemical characteristics
    of adrenergic receptors and pathways in brown-adipocytes, Proc.
    2nd Symp. Pharmacol. Thermoregul., Paris.
Horwitz, B. A., 1979, Cellular events underlying catecholamine-
    induced thermogenesis cation transport in brown adipocytes,
    Fed. Proc., 38(8):2170-2176.
Jansky, L., 1963, Body organ cytochromic oxidase activity in cold
    and warm acclimated rats, Can. J. Biochem. Physiol., 41:1847-
    1854.
Jansky, L., 1966, Body organ thermogenesis of the rat during exposure
    to cold and at maximal metabolic rate, Fed. Proc., 25:1297-
    1302.
Jansky, L., 1973, Non-shivering thermogenesis and its thermoregula-
    tory significance, Biol. Rev., 48:85-132.
Jansky, L., 1976, Effect of cold and exercise on energy metabolism
    in small mammals, in: "Progress in animal biometeorology," E.
    Tromps, ed., Swets and Zeitlinger Publ. Co., Amsterdam.

Kališnik, M., Vraspir-Porenta, O., Kham-Lindtner, T., Logonder-
    Mlinšek, M., Škrk, J., and Pajtnar, M., 1981, Stress and
    radiation carcinogenesis in mice, Neoplasma, 28:3.
Lowry, O. H., Rosebrough, N. J., Farr, A. L., and Randall, R., 1951,
    Protein measurement with the folin phenol reagent, J. Biol.
    Chem., 193:265-275.
McCord, J. M., 1979, Superoxide, superoxide dismutase and oxygen
    toxicity, Rev. Biochem. Toxicol.:109-124.
Misra, H. P., and Fridovich, I., 1972, The role of superoxide anion
    in the autooxidation of epinephrine and a simple assay for
    superoxide dismutase, J. Biol. Chem., 247:3170-3175.
Mitsuyuki, A., MD., Takehiro, N., MD., Yutaka, Y., MD., Masay, T.,
    MD., Koji, O., MD., Masahiro, H., MD., and Mariyoshi, R., MD.,
    1981, Studies on the radioprotective effects of superoxide
    dismutase in mice, Int. J. Radiat. Oncol. Biol. Phys., 7:
    205-209.
Petkau, A., and Chelack, W. S., 1976, Radioprotective effect of
    superoxide dismutase on model phospholipid membranes, Biochim.
    Biophys. Acta, 433:445-456.
Petkau, A., Chelack, W. A., and Pleskach, S. D., 1978, Protection
    by superoxide dismutase of white blood cells in x-irradiated
    mice, Life Sci., 22:867-882.
Petkau, A., 1980, Radiation carcinogenesis from a membrane perspec-
    tive, Acta Physiol. Scand. Suppl., 492:81-90.
Petkau, A., 1980, Radiation protection by superoxide dismutase,
    Photochem. Photobiol., 28:765-774.
Petrović, V. M., 1976, Effect of acute and chronic exposure to cold
    on the endocrine system of laboratory animals, in: "Progress
    in animal biometeorology," E. Tropms, ed., Swets and Zeitlinger
    Publ. Co., Amsterdam.
Petrović, V. M., Rajčić, O., Vrbaški, M., and Milić, B., 1979,
    Adaptation to cold and liver protein synthesis in the rat,
    Bull. Acad. Serbe Sci., 18:19-25.
Petrović, V. M., Saičić, Z., Spasić, M., Radojičić, R., and Milić,
    B., 1980, Effect of noradrenaline and thyroid hormones on
    superoxide dismutase activity in the liver, brown adipose
    tissue and brain, Bull. Acad. Serve Sci., 20:11-18.
Petrović, V..M., Gudz, T., and Saičić, Z., 1981, Selective effect
    of noradrenaline on superoxide dismutase activity in the brown
    adipose tissue and liver of the rat, Experientia, 37:14.
Petrović, V. M., Spasić, M., Saičiʹ Z., Milić, B., and Radojičić,
    R., 1982, Increase in superoxide dismutase activity induced
    by thyroid hormones in brains on neonate and adult rats,
    Experientia (in press).
Schwarz, H. P., Dreisbach, L., and Kleschnik, A., 1963, The effect
    of x-ray irradiation on the phospholipids of liver mitochondria,
    Arch. Biochem. Biophys., 101:103-107.
Wallach, D. F. H., and Widenkamm, E., 1973, Radiation effects in
    biomembranes, Klin. Wschr., 51:419-430.

ADAPTATIONS OF THE REPRODUCTIVE SYSTEM OF RODENTS TO CHANGING PHOTOPERIODIC CONDITIONS*

Russel J. Reiter

Department of Anatomy
University of Texas Health Science Center
San Antonio, Texas 78284, U.S.A.

INTRODUCTION

All mammals which inhabit the temperate and polar regions of the earth are seasonal breeders. Thus, they are capable of successful reproduction for only a finite period of time each year. The fluctuation in reproductive competence ensures that the young are born during a very restricted period of time. Most commonly this is the spring and early summer. Mammals have evolved mechanisms to ensure seasonal breeding for a very specific reason, i.e., it increases the chances of survival for the young. The warm days of the spring and summer in combination with increased food availability at this time of the year promotes survival of the newborns. Mammals have evolved a variety of different mechanisms to ensure seasonal reproduction. Many of them depend on the changing environmental photoperiod as a predictor of the upcoming season. These species are described as being photoperiodic.

A number of rodents are photoperiodic. They adjust their reproductive potential according to the prevailing light:dark cycle (Reiter, 1974). The following paragraphs will briefly summarize the mechanisms whereby the light:dark cycle determines the functional status of the reproductive system of these rodents.

CONVERSION OF PHOTOPERIODIC INFORMATION INTO A HORMONAL SIGNAL

In mammals the exclusive detectors of photoperiodic information

---

*Supported by NSF grant no. PCM 8003441

are the retinas of the lateral eyes. However, in addition to the
retinas being related to the visual and visual association areas of
the central nervous system, photic information is also transferred
to an organ on the posterodorsal aspect of the brain, the pineal
gland, where it is converted into a hormonal message (Fig. 1).

Neural Pathways which Connect the Retinas to the Pineal Gland

In the retinas the photoreceptor elements transduce photic
energy into a neural impulse which passes through the retinas and
eventually to the ganglion cells. The axons of these cells project
from the eyes to the suprachiasmatic nuclei of the hypothalamus.
In order to reach the pineal gland the neural information is transferred over yet undefined pathways through the hypothalamus and
eventually to the intermediolateral cell column of the upper
thoracic cord. From here the impulses are carried into the peripheral
autonomic nervous system to the superior cervical ganglia where the
final synapse is made (Fig. 1). Postganglionic fibers carry the
neural impulses to the pineal gland where they end in the vicinity
of pinealocytes, the functional units of the pineal gland. The
fibers end near of the pinealocytes but, at least on a morphological
basis, they do not form true synapses with the effector cells. The
central portions of this pathway have been primarily described by
Moore (1978) while Kappers (1960) defined the innervation of the
pineal gland via the peripheral sympathetic nervous system.

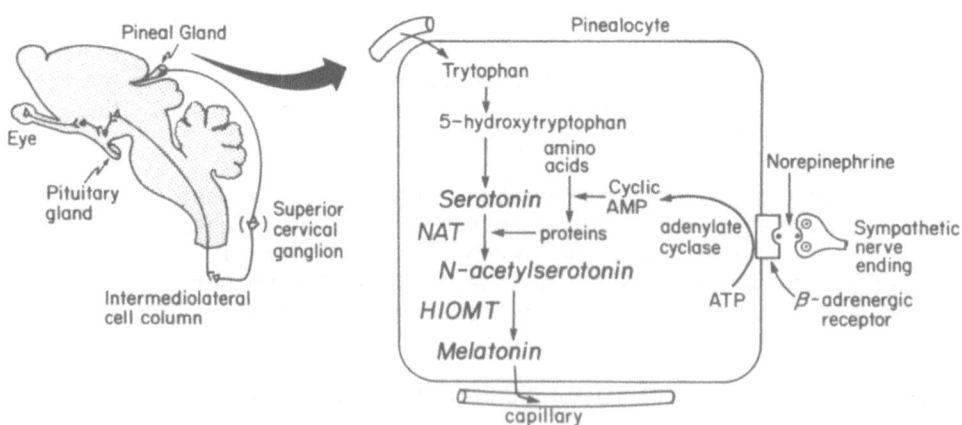

Fig. 1. Neural pathways which connect the eye to the pineal gland
        (left). On the right is represented the synthesis of mela-
        tonin in a pinealocyte, the functional unit of the pineal
        gland. NAT, N-acetyltransferase; HIOMT, hydroxyindole-o-
        methyltransferase.

Transduction of Neural Signal into a Chemical Output in the Pineal Gland

The nerve endings which terminate within the vicinity of the pinealocytes release, primarily during darkness, a neurotransmitter which acts on the cells to stimulate their synthetic machinery. Although the neurotransmitter is generally considered to be the norepinephrine(NE) based on studies using the rat (Zatz, 1981), this has yet to be proven for the hamster (Lipton et al., 1982), a species which has had great utility in studying the physiology of the pineal gland (Reiter, 1980). Once NE is released it acts on β-adrenergic receptors on the pinealocyte membrane; this stimulates a series of events in which cyclic AMP serves as a second messenger in the production of perhaps a number of pineal hormones (Oleshansky and Neff, 1978) (Fig. 1). The major hormonal product of the pineal gland seems to be melatonin (Reiter and Vaughan, 1977).

Melatonin is a product of tryptophan metabolism; this is an amino acid taken up by the pinealocytes from the blood (Axelrod, 1974) (Fig. 1). Tryptophan is converted, in a two step process, to serotonin, an indoleamine which serves as a common precursor of a number of pineal products. One of the major synthetic products of the pineal is melatonin (Klein et al., 1981). In the synthesis of melatonin, serotonin is first converted to N-acetylserotonin by the enzyme N-acetyltransferase (NAT). N-acetylserotonin is then O-methylated by the enzyme hydroxyindole-O-methyltransferase (HIOMT) with the resultant formation of N-acetyl-5-methoxytryptamine (melatonin).

The production of melatonin in the pineal gland is highly cyclic. The rhythm in melatonin production is closely correlated with the photoperiodic environment to which the animals are exposed. Thus, during the daily light period melatonin levels are characteristically low within the pineal gland; conversely, at night the synthesis of melatonin may increase many fold (Wurtman and Ozaki, 1978; Panke et al., 1979). Since melatonin seems not to be stored within the pineal gland, but rather is rapidly released into the blood vascular system, plasma concentrations of the indole follow very closely its production within the pineal gland (Wilkinson et al., 1977). Thus, plasma titers of the hormone are higher at night than during the day. Because of the marked rhythm in melatonin production and secretion from the pineal gland other organs can be apprised of the photoperiodic status.

It is generally believed that melatonin production is controlled primarily by the activity of the serotonin acetylating enzyme, i.e. NAT. When NAT activity is high, e.g., at night, melatonin levels are also elevated. The melatonin forming enzyme (HIOMT) seems to exhibit no rhythmic fluctuation and thus its importance in controlling melatonin production is considered minimal (Klein et al., 1981). Since melatonin is produced at one site (the pineal gland), is released into the blood vasculature, and travels to another organ system (the neuroendocrine-reproductive axis) to have its action, it meets all the criteria of a hormone.

Other potential pineal hormones include N-acetylserotonin, 5-methoxytryptophol, 5-methoxytryptamine, and various unidentified peptides (Brown et al., 1981; Reiter and Vaughan, 1977; Pevet et al., 1981; Benson and Ebels, 1981).

REPRODUCTIVE CONSEQUENCES OF THE PINEAL GLAND

When the pineal gland of a variety of rodent species is stimulated by darkness it eventually causes gonadal atrophy (Reiter, 1974, 1980). The animal that has provided the most information in this regard is the Syrian hamster (*Mesocricetus auratus*). It is a highly photosensitive species whose reproductive organs respond to changes in the daily light:dark cycle. For example, the exposure of either male or female hamsters to anything less than 12.5 hours of light daily causes the gonads of this species to undergo total involution (Elliott, 1976). This is known to be a pineal mediated response since if the gland is surgically removed or if it is sympathetically denervated, darkness is completely incapable of influencing the reproductive system (Hoffmann and Reiter, 1965; Reiter and Hester, 1966). The changes in the reproductive organs of intact hamsters exposed to short days are dramatic. In males, the testes regress to one-sixth their normal size, spermatogenesis is arrested, and reproductively-active hormones are depressed (Berndtson and Desjardins, 1974) (Fig. 2); in females, ovulation is interrupted, the uteri atrophy, vaginal cyclicity ceases, and the gonadotrophin levels are altered (Seegal and Goldman, 1975). Again, in pineal-ectomized hamsters these changes, normally induced by short day exposure, do not occur. Other rodents where the effects of the pineal gland on the reproductive organs are equally as dramatic include the Djungarian hamster (*Phodopus sungorus*) (Hoffmann, 1973), the white-footed mouse (*Peromyscus leucopus*) (Lynch, 1973) and the vole (*Microtis agrestis*) (Charlton et al., 1976).

Since melatonin is the prime candidate for being the essential pineal antigonadotrophic agent of the pineal gland in the hamster, its administration to animals kept in long days should cause gonadal regression. Although it took a number of years to define the proper experimental paradigm, it is now known that properly timed melatonin injections (daily injections late in the light period) are capable of inducing the same degree of atrophy as does short day exposure (Tamarkin et al., 1976) (Fig. 2). No other pineal compound has yet been shown to have equivalent inhibitory effects on the reproductive system of hamsters.

The potent inhibitory signal which darkness, acting by the way

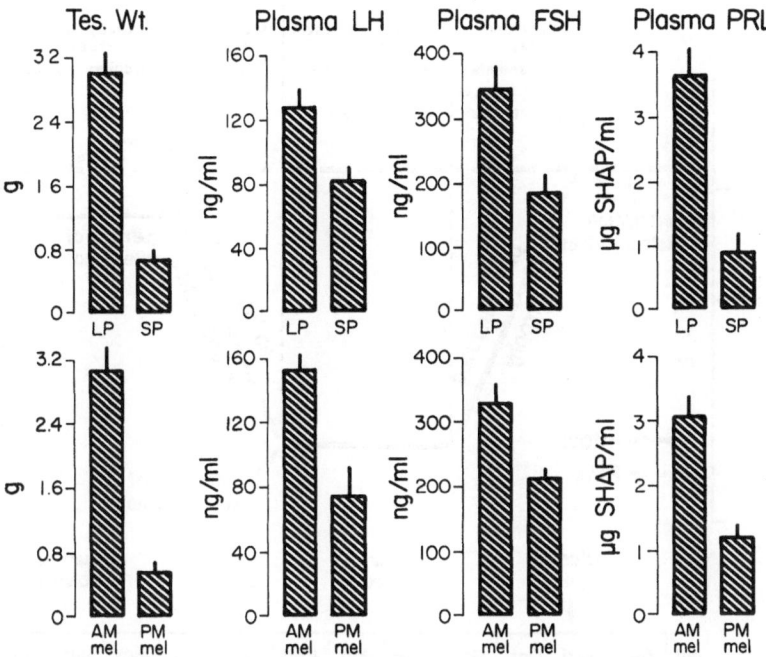

Fig. 2. Testicular weights (Tes. Wt.) and plasma levels of lutein-
izing hormone (LH), follicle stimulating hormone (FSH), and
prolactin (PRL) in male hamsters kept under either long
(LP) or short photoperiods (SP). Also shown are the effects
of morning (AM mel) or afternoon (PM mel) melatonin injections on these reproductive parameters. Daily afternoon injections of melatonin induce the same degree of reproductive change as does short day exposure.

of the pineal gland, exerts on the reproductive system of these
species is especially important to these animals when they are in
the field (Reiter, 1974). Thus, as day-lengths decrease in the fall
of the year the pineal gland becomes maximally stimulated and
gonadal atrophy follows. These relationships have been especially
well studied in the Syrian hamster. The annual cycle of reproduction has been divided into four clearly defined phases, i.e., the
inhibition phase, the sexually quiescent phase, the restoration
phase, and the sexually active phase (Reiter, 1975a) (Fig. 3).

The decreasing functional state of the reproductive organs
during the period of shorter days in the fall of the year is referred
to as the inhibition phase. The changes seem in the neuroendocrine-
reproductive system in these animals are the same as those which
follow the exposure of hamsters to short days in the laboratory
(Reiter, 1978). The length of time required for the gonads to
completely involute is roughly 6-8 weeks for the Syrian hamster.

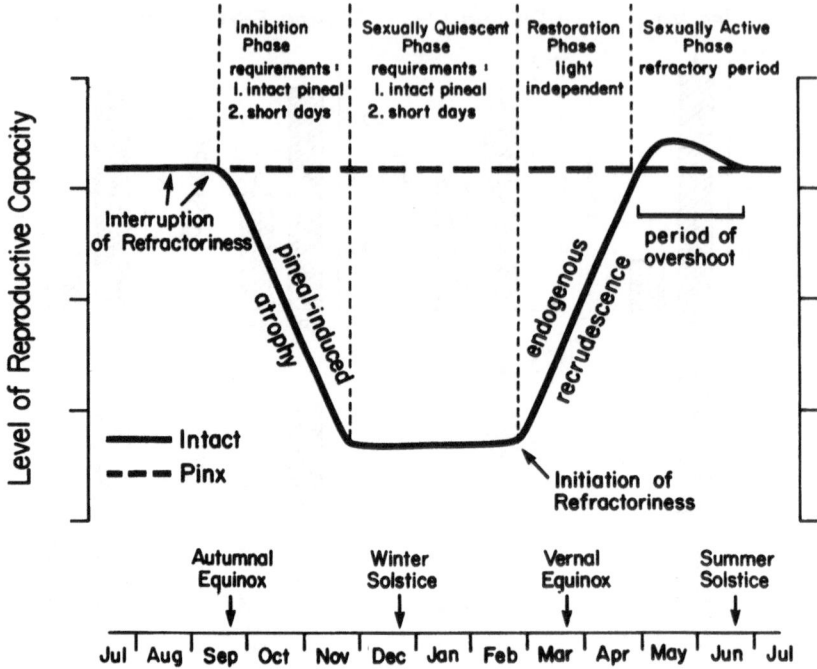

Fig. 3. Changes in the reproductive capacity during the various
phases of the annual reproductive cycle. These data were
primarily garnered from studies in which the Syrian hamster
was the experimental animal.

At the end of this phase the animals are reproductively incompetent.
They then enter the sexually quiescent phase. This phase of the
annual cycle is rather prolonged, extending from early fall to well
after the winter solstice. The sexual quienscence imposed by the
short days of the fall and winter ensure that the animals do not
breed and deliver young during times of the year which would not
be maximally conducive to the survival of the newborns. Additionally, during this phase hamsters are usually confined to dark underground burrows where they presumably hibernate; the atrophic reproductive organs also promote hibernation (Hall and Goldman, 1980).

As spring approaches, but before hamsters emerge from hibernation, i.e., while they are still in darkness, their gonads begin
to regrow; this initiates the restoration phase (Fig. 3). The re-
growth of the sexual organs apparently occurs because the neuroendocrine-reproductive system becomes refractory to the inhibitory
influence of the pineal gland (Reiter, 1975b), rather than to a
discontinuation of pineal function. The recrudescence of the reproductive organs requires about 8 weeks. During this interval

gametogenesis is restored in both sexes and hormonal patterns return
to normal. At the end of this phase the animals emerge above ground
in the spring.

The sexually quiescent phase extends from early spring until
the subsequent fall when the entire cycle is again initiated. During
the sexually active period each female probably has several litters
of young. The refractoriness that developed to initiate gonadal restoration is interrupted by the long days of the summer (Stetson
et al., 1976).

The seasonal cycle described in the preceding paragraphs is
carefully synchronized with the appropriate seasons by the changing
photoperiod acting by way of the pineal gland. When animals are
pinealectomized their reproductive organs remain functional even
during the winter months (Reiter, 1973/74) (Fig. 2). This is not
an advantage since young born during the winter would likely die
because of the harsh environmental conditions and the limited food
supplies. Thus, the pineal gland, since it generates the seasonal
reproductive cycle, is probably critical to the survival of the
species in its natural habitat.

CONCLUDING REMARKS

It is clear that in some species the photoperiod is critical
in determining the reproductive state of the animal. This ensures
that these animals are seasonal breeders under natural photoperiodic
conditions. The pineal gland serves to transduce neural information
into a chemical signal which controls reproduction. This chemical
signal is most likely melatonin.

REFERENCES

Axelrod, J., 1974, The pineal gland: A neurochemical transducer,
    Science, 184:1341-1348.
Benson, B., and Ebels, I., 1981, Other pineal peptides and related
    substances - Physiological implications for reproductive bio-
    logy, in:"The pineal gland. Vol. II. Reproductive effects,"
    R. J. Reiter, ed., CRC Press, Boca Raton.
Berndtson, W. E., and Desjardins, C., 1974, Circulating LH and FSH
    levels and testicular function in hamsters during light de-
    privation and subsequent photoperiodic stimulation, Endocrino-
    logy, 95:195-204.
Brown, G. M., Tsui, H. W., Niles, L. P., and Grota, L. J., 1981,
    Gonadal effects of the pineal gland, in:"Pineal function,"
    C.D. Matthews, and R.F. Seamark, eds., Elsevier, Amsterdam,
    North Holland.
Charlton, H. M., Grocock, C. A., and Ostberg, A., 1976, The effects

of pinealectomy and superior cervical ganglionectomy on the testis of the vole, *Microtus agrestis*, J. Reprod. Fertil., 48: 377-379.

Elliott, J., 1976, Circadian rhythms and photoperiodic time measurement in mammals, Fed. Proc., 35:2339-2346.

Hall, V., and Goldman, B., 1980, Effects of gonadal steroid hormones on hibernation in the Turkish hamster (*Mesocricetus brandti*), J. Comp. Physiol., 135:107-114.

Hoffmann, K., 1973, The influence of photoperiod and melatonin on testis size, body weight, and pelage colour in the Djungarian hamster (*Phodopus sungorus*), J. Comp. Physiol., 95:267-282.

Kappers, J. A., 1960, The development, topographical relations and innervation of the epiphysis cerebri in the albino rat, Z. Zellforsch., 52:163-215.

Klein, D. C., Auerbach, D. A., Namboodiri, M. A. A., and Wheler, G. H. T., 1981, Indole metabolism in the mammalian pineal gland, in:"The pineal gland. Vol I. Anatomy and biochemistry,", R.J. Reiter, ed., CRC Press, Boca Raton.

Lipton, J. S., Petterborg, L. J., Steinlechner, S., and Reiter, R. J., 1982, In vivo responses of the pineal gland of the Syrian hamster to isoproterenol or norepinephrine, in:"The pineal gland;its hormones," R.J. Reiter, ed., Alan R. Liss, New York.

Lynch, G. R., 1973, Effect of simultaneous exposure to differences in photoperiod and temperature on the seasonal molt and reproductive system of the white-footed mouse, *Peromyscus leucopus*, Comp. Biochem. Physiol., 53C:67-68.

Moore, R. Y., 1978, The innervation of the mammalian pineal gland, in:"The pineal and reproduction," R.J. Reiter, ed., Karger, Basel.

Oleshansky, M. A., and Neff, N. H., 1978, Studies on the control of pineal indole synthesis: Cyclic nucleotides, adenylate cyclase and phosphodiesterase, J. Neural Transmis. Suppl., 13:81-96.

Panke, E. S., Rollag, M. D., and Reiter, R. J., 1979, Pineal melatonin concentrations in the Syrian hamster, Endocrinology, 104: 194-197.

Pevet, R., Haldar-Misra, C., and Öcal, T., 1981, The independency of an intact pineal gland of the inhibition by 5-methoxy-tryptamine of the reproductive organs in the male hamster, J. Neural Transmis., 52:95-106.

Reiter, R. J., 1974, Circannual reproductive rhythms in mammals related to photoperiod and pineal function: A review, Chronobiologia, 1:365-395.

Reiter, R. J., 1973/74, Influence of pinealectomy on the breeding capability of hamsters maintained under natural photoperiodic and temperature conditions, Neuroendocrinology, 13:366-370.

Reiter, R. J., 1975a, The pineal gland and seasonal reproductive adjustments, Int. J. Biometeorol.,19:282-288.

Reiter, R. J., 1975b, Evidence for refractoriness of the pituitary-gonadal axis to the pineal gland in golden hamsters and its possible implications in annual reproductive rhythms, Anat. Rec., 173:365-372.

Reiter, R. J., 1978, Interaction of photoperiod, pineal and seasonal reproduction as exemplified by findings in the hamster, in: "The pineal and reproduction," R.J. Reiter, ed., Karger, Basel.

Reiter, R. J., 1980, The pineal and its hormones in the control of reproduction in mammals, Endocr. Rev., 1:109-131.

Reiter, R. J., and Hester, R. J., 1966, Interrelationships of the pineal gland, the superior cervical ganglia and the photoperiod in the regulation of the endocrine systems of hamsters, Endocrinology, 79:1168-1170.

Reiter, R. J., and Vaughan, M. K., 1977, Pineal antigonadotrophic substances: Polypeptides and indoles, Life Sci., 21:159-172.

Seegal, R. F., and Goldman, B. D., 1975, Effects of photoperiod on cyclicity and serum gonadotropins in the Syrian hamster, Biol. Reprod., 12;223-231.

Stetson, M. H., Matt, K. S., and Watson-Whitmyre, M., 1976, Photoperiodism and reproduction in golden hamsters: Circadian organization and the termination of photorefractoriness, Biol. Reprod., 14:531-539.

Tamarkin, L., Westrom, W. K., Hamill, A. I., and Goldman, B. D., 1976, Effect of melatonin on the reproductive systems of male and female Syrian hamsters: A diurnal rhythm in sensitivity to melatonin, Endocrinology, 99:1534-1541.

Wilkinson, M., Arendt, J., Bradtke, J., and de Ziegler, D., 1977, Determination of a dark-induced increase of pineal N-acetyl transferase activity and simultaneous radioimmunoassay of melatonin in pineal, serum and pituitary tissue of the male rat, J. Endocr., 72:243-244.

Wurtman, R. J., and Ozaki, Y., 1978, Physiological control of melatonin synthesis and secretion: Mechanisms generating rhythms in melatonin, methoxytryptophol, and arginine vasotocin levels and effects on the pineal of endogenous catecholamines, the estrous cycle, and environmental lighting, J. Neural Transmis. Suppl., 13:59-70.

Zatz, M., 1981, Pharmacology of the rat pineal gland, in:"The pineal gland. Vol. I. Anatomy and biochemistry," R.J. Reiter, ed., CRC Press, Boca Raton.

ENDOCRINE CYCLES AND HIBERNATION IN THE HEDGEHOG: MECHANISMS OF ADAPTATION TO NATURAL VARIATIONS IN THE ENVIRONMENT

M. Saboureau and J. Boissin

Centre d'Etudes Biologiques des Animaux Sauvages
C.N.R.S.
Villiers-en-Bois, 79360 Beauvoir-sur-Niort, France

INTRODUCTION

The hedgehog is a hibernating mammal with a fairly wide geo-graphical distribution in Europe ranging from the arctic circle to the 40th North parallel and consequently is an interesting choice for the study of the adaptation of the individual animal and of the species as a whole to seasonal variations in the environment. Hibernation is a complex process which ensures the survival of the individual hedgehog during the cold season in autumn and winter. In the hedgehog the duration of the hibernating period is variable depending on latitude and climate (Herter, 1933; Kristoffersson and Soivio, 1967; Johansson and Senturia, 1972; Walhovd, 1973, 1975; Morris, 1973; Berthoud, 1978; Saboureau, 1979). In addition, during hibernation, there is adaptation or modification in the functions of the numerous endocrine glands involved in the regulation of this phenomenon. Former studies on numerous hibernating mammals have shown that winter lethargy does not exist unless there is involution of the endocrine glands; it must be stressed however, that we can no longer adhere to this point of view, even if all evidence of hibernation is obviously under the control of multiple hormonal regulations (Musacchia and Jansky, 1981). Thus it was interesting to study the adaptation of the hedgehog (*Erinaceus europaeus* L.) to seasonal variations in the environment, in the west of France, one of the southern most regions in Europe where this species can be found. During the present study, the annual cycle of general locomotor activity and the phenomenon of hiber-nation were determined and in addition concomitant seasonal processes in endocrine-metabolic activities (Cortico-adrenal and thyroid functions) and genital activities were characterized.

MATERIALS AND METHODS

Animals

Adult male hedgehogs were captured in the Deux-Sèvres district (France). They were bred in individual 6 m^2 parks situated on the limits of the Chizé Forest (latitude 46° 07' N; longitude 0° 25' W) and exposed to natural seasonal climatic variations (light, temperature and rainfall). The animals were fed daily with a mixture of crushed chicken meat and dog biscuits (Canina, Duquesne-Purina). Straw was provided in each pen for the animal's nest, and the tunnel giving access to the nest was equipped with a system of detection composed of a double beam of infrared rays. This system was linked to a graphic recorder (Polystyl-Schlumberger) which enabled us to determine the direction of passage, the number of entries and exits, and their situation in time (Fig. 1). In these conditions, the animals presented characteristics comparable to those of wild hedgehogs.

Methods

Each month blood samples were taken from the jugular vein after anaesthesia with fluothane (ICI Pharma). The heparinized blood was centrifuged and each sample was divided into several aliquots and

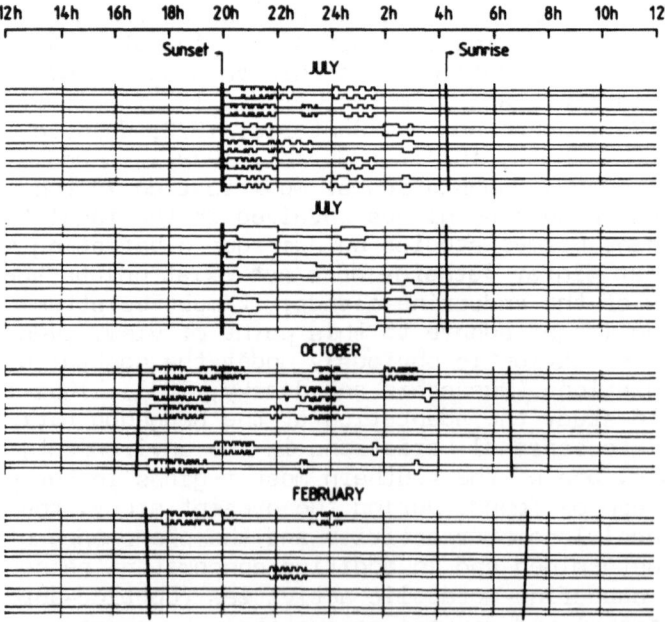

Fig. 1. Situation, during 24 hours periods and at different times of the year, of nest entries and exits in the hedgehog.

immediately stored at -25°C until assay. In addition, in order to
determine the seasonal cycle of the cortico-adrenal gland, blood
samples were taken each month, every four hours during 24 hours,
from non-anaesthetized active or resting animals equipped with an
implanted intra-arterial catheter (Saboureau et al., 1980).

Because of the important and constant ratio cortisol/corticos-
terone (F/B=28) in the hedgehog (Boissin et al., 1975a), the two
hormones were assayed without chromatographic separation by a
competitive protein binding technique using human plasma as source
of transcortine, after elimination of the endogen steroids
(Saboureau et al., 1979). Plasma thyroxine (T4) was assayed by a
radio-competition technique described by Vigouroux (1972). Testicular
activity was determined by measuring the dimensions of the testes
after laparotomy and estimation of testicular volume and by the
determination of plasma testosterone levels (T) using a radioimmunoassay (^{3}H-Testosterone RIA, Biomerieux, Fr.) (Saboureau and
Dutourné, 1981). The production rates of testosterone (PR) were
calculated from plasma testosterone concentrations (c) and the
metabolic clearance rates (MCR), determined by the single injection
technique (Saboureau, 1979), according to the formula:

$$PR = MCR \times (c)$$

Statistical Analysis

All the data were expressed as means $\pm$ the standard error (SEM),
and statistical comparison between the means were evaluated using
the analysis of variance (F-test).

RESULTS

Adaptation of the Animal

Seasonal variations in locomotor activity. The data recorded
at different times of the year (Fig. 1) show that the locomotor
activity of hedgehogs kept in parks is strictly nocturnal, similar
to that of wild hedgehogs. Examination of the activity periods shows
that the animals are more active during the first part of the night.
In autumn and in winter, activity phases are of short duration and
are not synchronized with each other, which leads us to think that
they correspond to phases of spontaneous arousals; they are separated by long periods of inactivity which have been assimilated
to phases of hypothermic lethargy (torpor). The alternation of these
different sequences constitutes the period of hibernation. The
daily position of the activity period (i.e. the time elapsed starting from the first exit and ending with the last entry) (Fig. 2)
shows that general locomotor activity, variable for each animal, is
reduced in autumn and during early winter, whereas from February to

the end of summer all the animals are active every night. During
this period, the beginning of the activity sequence is synchronized
with sunset and the first exit from the nest occurs less than an
hour after sunset whereas activity stops about an hour before sunrise. In autumn and in winter, the activity sequences, often of
short duration and always situated during the night, are no longer
synchronized with sunset.

The first resting sequences can be observed at the end of
summer (Fig. 3) and are not related to any modification in climatic
conditions; in autumn, they become more and more frequent and their
duration gradually increases as the outer temperature decreases.
The duration of the torpor phases the most frequently observed does
not exceed 4 to 6 days whereas the activity phases are short (1 or
2 days). In the west of France, the end of hibernation in the adult
male hedgehogs usually occurs in February-March. During our study,

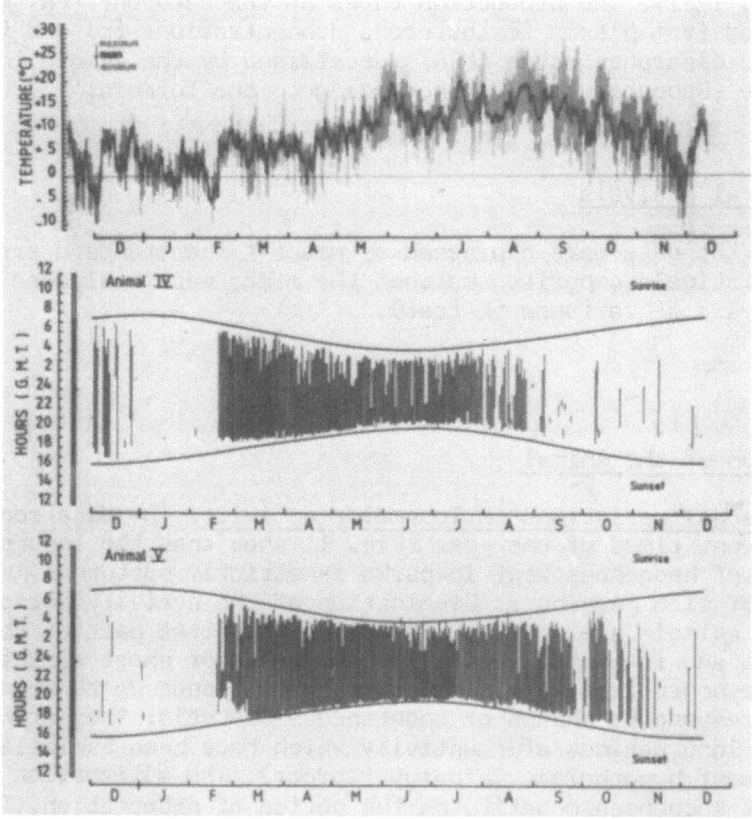

Fig. 2. Seasonal variations of daily external temperature and daily
locomotor activity period in the hedgehog. The vertical
lines, between sunset and sunrise, represent the activity
of the animal from the first exit to the last entry into
the nest.

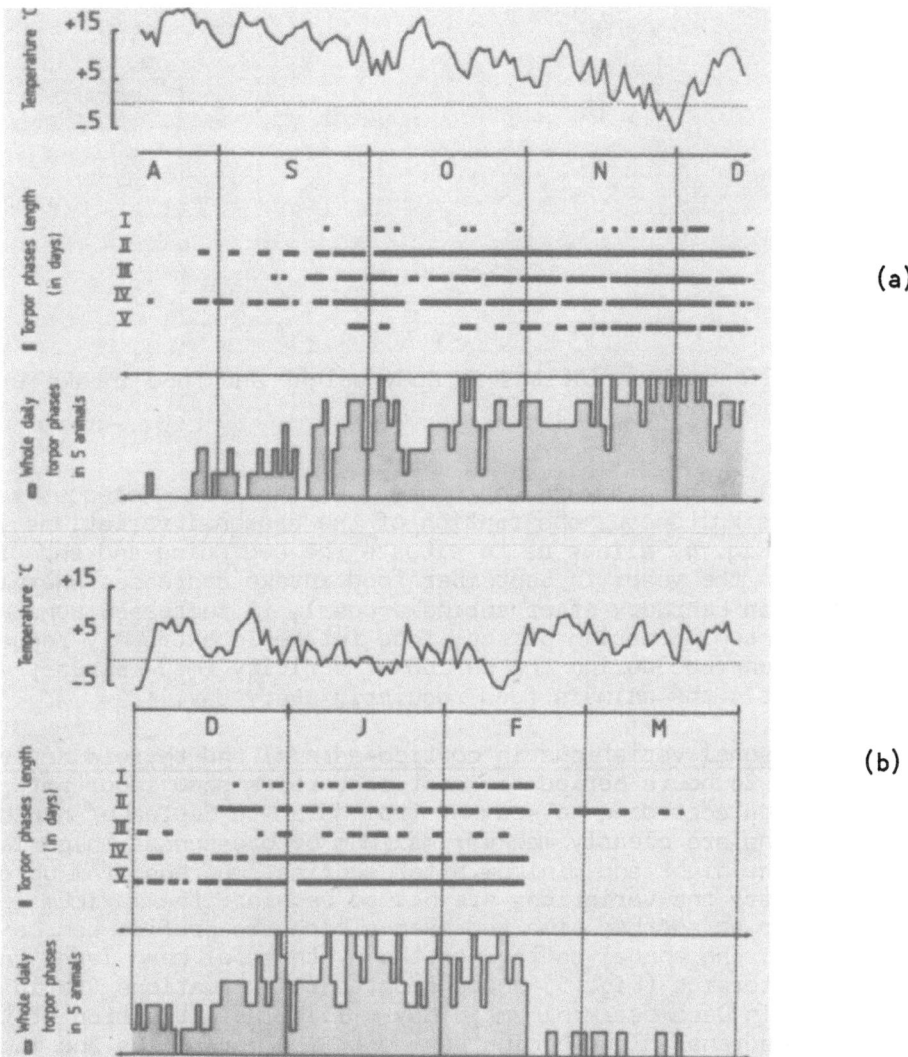

Fig. 3. Distribution of torpor and activity sequences in the hedge-
        hog during the hibernating period: a) beginning of hiber-
        nation; b) end of hibernation.

the spring arousal took place suddenly in mid-February and was concomitant with a considerable increase in ambient temperature which
in 48 hours rose from -5.5°C to +8°C.

Seasonal variations in body weight. The seasonal body weight
cycle is clearly defined (Fig. 4) and is characteristic of hibernating mammals. Body weight is minimum in winter, followed by an
increase in summer and reaches a maximum level in early autumn

Fig. 4. Seasonal variations in body weight and food intake in the hedgehog.

(October). During autumn and winter, weight then decreases regularly. The simultaneous observation of the seasonal variations in food
intake (Fig. 4) allows us to situate the beginning and end of hibernation in the year: in September food intake decreases sharply
whereas in February after spring arousal, it increases considerably.
During the hibernating period, food intake is extremely reduced
(< 50%) whereas during the period of activity (mid-February - end
of August), the animals feed regularly every day.

Seasonal variations in cortico-adrenal and thyroid activities.
During a 24 hours period the corticosteroids show important
variations according to season: from March to September nycthemeral
variations are clearly marked: maximum before sunset, high levels
during the night and minimum after sunrise, whereas from October
to February the variations are not so regular: the maximum level
may occur at another time and even a biphasic profile can be
observed. The annual pattern of the nycthemeral mean level in
corticosteroids (Fig. 5) shows important fluctuations (ratio
maximum in December/minimum in May = 13). The activation of the
cortico-adrenal gland begins during the summer months and increases
in autumn at the beginning of the hibernating period. The maximum
is reached in December (130.2 ± 4.8 ng/ml) but hormone levels then
decrease very rapidly (January: 48.0 ± 5.8 ng/ml; December *vs* January:
P < 0.01) until February (19.1 ± 1.8 ng/ml). Although a slight increase occurs in March and April, plasma cortico-adrenal hormone
level decreases in May (9.8 ± 2.0 ng/ml). Recrudescence of the cortico-adrenal activity takes place in June. During summer and autumn
the mean nycthemeral level rises regularly in spite of a slight
but not significant decrease in September.

The thyroid activity cycle analysed by the variations in
plasma thyroxine levels (T4) also shows a distinct annual pattern
(Fig. 6). Minimum values are observed during autumn (6-7 ng/ml).
The thyroid resume its activity in winter and T4 levels are high

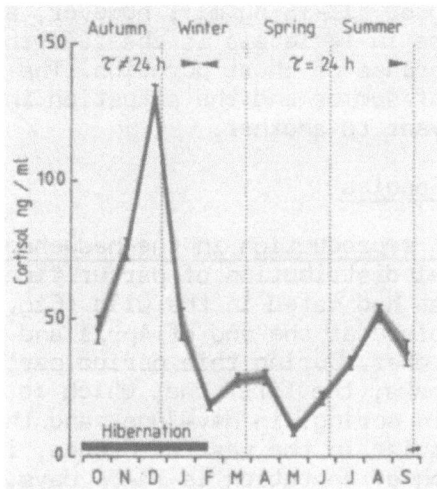

Fig. 5. Monthly nycthemeral mean level variations of plasma corticosteroids in the hedgehog.

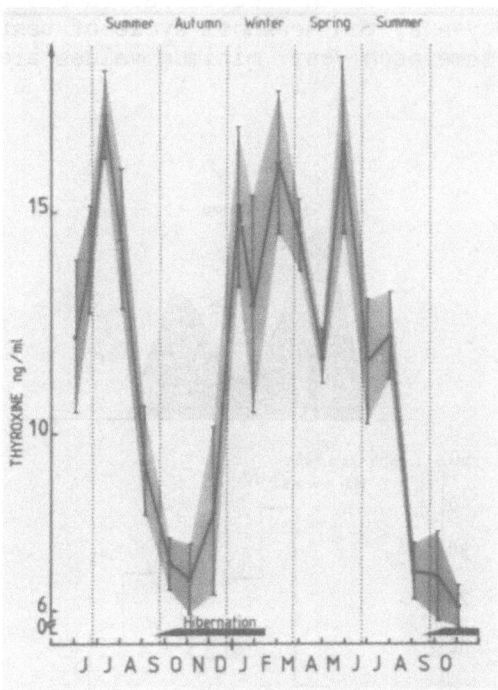

Fig. 6. Annual cycle of plasma thyroxine levels in the hedgehog.

from January to August (12-16 ng/ml); however, a statistically
significant decrease in T4 levels is observed in May, followed in
June by a sharp increase of short duration. The fall in thyroxinemia
occurs at the end of summer and the situation in September can be
variable from one year to another.

Adaptation of the Species

The biology of reproduction in the hedgehog in mid-western
France. The seasonal distribution of parturitions in captivity involving females that had mated in the wild (Fig. 7) shows that the
first births take place at the end of April and the latest births
are observed in October. During this period parturition frequencies
show two maximum peaks; the first one, which is also the most
important, occurs in spring, in May-June, and the second of less
importance, but regular in the west of France, in September-October.
As the mean duration of gestation is 34-36 days, we can therefore
deduce the rutting periods. Thus in the west of France, there are
two main rutting periods: from the end of March to the end of May
and from the end of July to the beginning of September, with a
decrease in mating in June and July.

Seasonal variations in the testicular function. Studied during
three consecutive years, the seasonal cycle of testicular volume
(Fig. 8) was the same each year: minimum values are obtained from

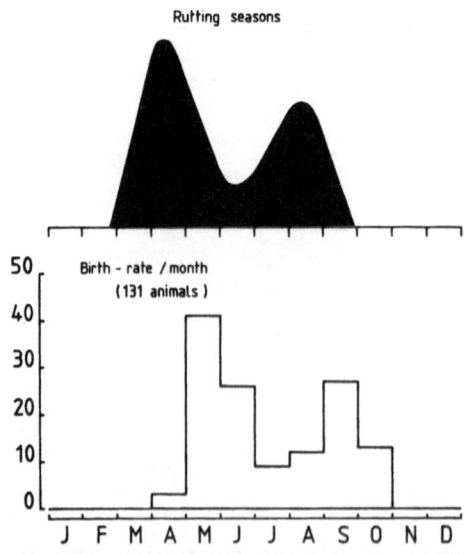

Fig. 7. Distribution during the year of rutting periods and births
in hedgehogs living in the west of France.

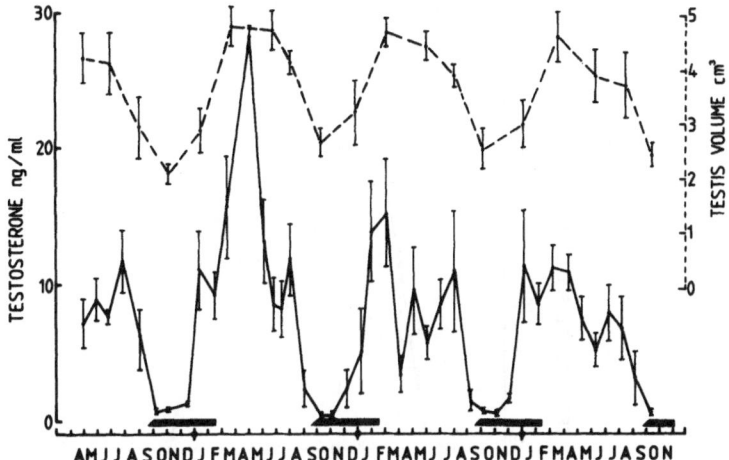

Fig. 8. Seasonal variations in testicular volume and plasma testosterone levels in the hedgehog living in the west of France.
Study carried out on the same group of animals during
several years.

October to December, activity is resumed in January-February, testicular volume is maximum from March to August and involution occurs
at the end of summer. Seasonal variations in plasma testosterone
levels follow a regular succession of characteristic sequences every
year although hormone levels may be different from one year to another. The quiescent phase which occurs in autumn (September-October
to December) is characterized by very low levels of testosterone for
all animals (0.4-1.0 ng/ml) although sometime testosterone levels
begin to rise as early as December. Testicular activity is resumed
in winter, during the month of January, with considerable individual
variability and variable intensity from one year to another. The
period of full testicular activity occurs between February and August
when high levels of testosterone are observed. It is characterized
by a biphasic profile with an important spring peak (10 to 15 ng/ml),
a significant regular decrease in June and a second peak in July-
August; these last plasma testosterone levels are slightly higher th
than in June but are not statistically different. In the west of
France these two peaks correspond to the two rutting periods observed.
Involution begins at the end of summer during the month of September.

The testosterone production rate (Fig. 9) shows that the endocrine function of the testis is quiescent in autumn, resumes activity
in January but with considerable individual variations and is at a
maximum in February-March. In April the production rate decreases
significantly and is low in June, before increasing again in July
and August. Thus, one of the essential characteristics of the endocrine activity of the testis is the resumption of activity during
winter, so testosterone levels are maximum at the end of hibernation.

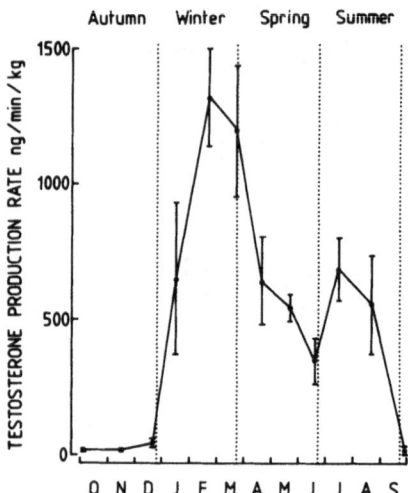

Fig. 9. Annual cycle in the testosterone production rate by the
testis in the hedgehog.

DISCUSSION

    The study of seasonal variations in generalized locomotor ac-
tivity in adult male hedgehogs living in the west of France allows
us to distinguish two periods: the active period (February-March
September) and the hibernating period (September to mid-February).
The beginning of the latter seems to be independent of outer
temperature and, in the west of France, take place about a month
earlier than in European regions further to the North where the
reduction in locomotor activity is not observed until October-
November (Saure, 1969; Saboureau et al., 1979). The precocity of
hibernation observed in the west of France could correspond to a
higher level of sensitivity to temperature at the end of summer
in the animals living in the region farthest to the South, as
Kristoffersson and Soivio (1967) observed, when they placed
hedgehogs from Germany and Finland at a temperature of +4°C at the
same time of the year (end of summer). The end of hibernation in
the hedgehog varies in Europe in relation to latitude and climate
(Saure, 1969; Saboureau et al., 1979); going from the south to the
north of Europe, hibernation occurs from the end of February (west
of France) to the end of April (Finland). The mild oceanic temperate
climate in the west of France seems to be one of the causes of this
extremely precocious spring arousal. These results are not incompatible with the hypothesis described by Pengelley and Asmundson
(1974) whereby hibernation is a rhythmic endogenous phenomenon in
which environmental factors play the role of synchronizer of this
internal clock. In this case we also have to admit that the
sensitivity of hedgehogs to changes in temperature can vary in

relation to their adaptation to different climatic conditions existing in Europe.

The results concerning the cortico-adrenal function in the hedgehog constitute the first evidence of the existence of a nyc-themeral rhythm of cortico-adrenal hormone, from February to October, in a wild animal (Saboureau et al., 1979). This rhythm, already described in numerous laboratory animals (Boissin et al., 1975b) can be compared to the rhythm observed in animals exhibiting noc-turnal activity (Szafarczyk et al., 1975). In animals that are active during the night, the corticosteroid levels are higher in the evening whereas for animals which exhibit the opposite behavior the highest values are observed in the morning. This endocrine rhythm is perfectly adapted to behavior needs during the passage from rest to activity; the cellular metabolism is already heavily stimulated by the cortico-adrenal hormones at this time. The suppression of this rhythm during autumn is difficult to explain but the hypothesis of a free running activity of the cortico-adrenal gland during hibernation can be postulated. The most important cortico-adrenal activity is observed during the first part of hibernation. Furthermore, although large seasonal varia-tions have already been described, this is the first time that evidence of an annual cycle in plasma corticosteroids has been shown in the hedgehog. Suomalainen and Nyholm (1956) interpret the rise in this endocrine activity at the beginning of hibernation as a reaction to stress induced by the fall in outer temperatures and the scarcity of food. Boulouard (1967), in the garden mouce, suggests that the increase in cortico-adrenal activity in autumn corresponds to a struggle mechanism against the cold which is not in relation to outer temperature but rather in relation to the seasonal thermic gradient. Thus, in autumn the animals would register the successive decreases in environmental temperature which would consequently lead to the physiological preparation for hibernation. However, in the hedgehog, summer reactivation of cortico-adrenal function occurs without any important climatic change and the premonitory characteristic of this endocrino-meta-bolic regulation is not easy to explain unless we admit that the cortico-adrenal function is completely integrated in the complex neuro-endocrine regulation sequences gifted with an automatism which controls hibernation (Pengelley and Asmundson, 1974).

The thyroid gland presents a clearly defined annual cycle characterized by a decrease in T4 levels at the end of summer and during autumn, and by the resumption of activity at the end of hibernation. The numerous studies concerning the thyroid function in hibernating animals often present contradictory results depend-ing on the species (Hudson and Wang, 1979; Hudson, 1981). Our results are similar to those obtained in the garden dormouse (Lachiver et al, 1957; Gabe et al., 1963), in the European hamster (Petrović and Kayser, 1958), and in several species of American squirrels (Hoffman and Zarrow, 1958; Hulbert and Hudson, 1976). In the hedgehog, hypothyroid activity observed at the end of summer could be responsible for the increase in lipidic reserves, as shown by the increase in body weight at this time. Besides, Augee et al. (1979) have shown that reduction in thyroid activity is related to an increase in membrane fluidity. The reason for the increase in thyroxine secretion at the end of the hibernating period might be the mobilisation of glucid reserves in order to maintain the normo-glycemia which the basic cellular reactions require. However, it must be noted that the regulation of these metabolic processes does not depend on only one hormone and that an exhaustive analysis of this phenomenon would involve consideration of the cateholamines and pancreas hormones.

In numerous hibernating species, the reproductive cycle is similar to the cycle observed in most perfect homeothermic species. Besides, the winter activation of the endocrine structures of the testis as described in the hedgehog seems to be typical of hiber-nating animals in general (Wimsatt, 1969). From the point of view of adaptation, this activity of the testicular function during hibernation is a remarkable phenomenon since the reproductive period can begin as soon as the cold season ends. The beginning of the reproductive period thus depends on the time of spring arousal. In the west of France, hedgehogs are able to breed as early as March, but in northern latitudes the rutting season takes place later, and this determines the duration and the number of re-productive periods (Saure, 1969). In northern European countries (Finland, Sweden, the Baltic Region, Poland, North Germany) the reproductive season begins in April-May, the youngs are born in July-August and there is only one litter per year, whereas else-where in Europe (South Germany, Holland, England, France) the re-productive period continues from March-April to September with usually two litters each year (Saure, 1969; Saboureau and Dutourné, 1981). These observations show that climatic factors can influence the sexual function and the duration of reproduction in relation to latitude. This faculty of adaptation allows us to understand why this species has an area of Geographic distribution so vast, from the Arctic circle to the 40th North parallel. In hibernating mammals, published studies offer little information on the season-al cycle of testosterone levels and it is not clear how the gonado-tropic hormones stimulate the interstitial tissue during winter. Besides, the phenomenon is made all the more complex by the fact that the animals hibernate inside a shelter in darkness during winter, so it is difficult to relate the recrudescence of the re-productive function to variations in the environment. Dussart and Richoux (1973) think that in the Dormouse the reactivation of the testicular function might be the result of microstimulations (gonadotropic hormonal discharges) which take place during spon-taneous arousals. Whether this is the case or not, the increase in testosterone production levels in December and January reveals

the important neuroendocrine sollicitation of the gonads towards
the end of the hibernating period.

If the annual cycles of the cortico-adrenal, thyroid and testicular functions are compared, the most important event is the
winter recrudescence in endocrine activity in these three functions.
The phase relationships observed in these cycles reveal specific
interactions. Cortico-adrenal activation preceeds the recrudescence
of testicular activity; during the period of full testicular activity, cortisol levels are at a minimum and in summer when the testis
function decreases, the cortico-adrenal function increases. The
existence of simultaneous phase relationships of testicular and
thyroid functions during the greater part of the year is rarely
found in published studies where it is more generally admitted as
well in birds (Assenmacher et al., 1975) as in non-hibernating
mammals (Badger: Maurel et al., 1977; fox: Maurel and Boissin, 1981;
mink: Boissin-Agasse et al., 1981) that there are negative interactions which lead to opposite relationships between the two annual
endocrine cycles. In the hedgehog, opposition only occurs from May
to July. It is as yet difficult to determine the nature of the
phase relationships which link these three cycles and other studies
in this field offer much contradictory informations. However, whatever the importance and the place of the hormonal interactions and
retroactions, we can suggest the following hypotheses: either the
three seasonal endocrine cycles are dependent on different neuroendocrine nervous structures equally sensitive to external factors
or it may be that they are controlled by a common and fundamental
oscillatory mechanism which might include several efferent neuroendocrine pathways assuring the cyclic activity of the testis,
the cortico-adrenal and the thyroid functions, and consequently
would enable both the species and the animal to adapt perfectly to
changes in the environment.

REFERENCES

Augee, M. L., Raison, J. K., Hulbert, A. J., 1979, Seasonal changes
    in membrane lipid transitions and thyroid function in the
    hedgehog, Am. J. Physiol., 236:E 589-E 593.
Assenmacher, I., Astier, H., Daniel, J. Y., Jallageas, M., 1975,
    Experimental studies on the annual cycles of thyroid and
    adrenocortical functions in relation to the reproductive cycle
    of drakes, J. Physiol. (Paris) 70:507-520.
Berthoud, G., 1978, Note préliminaire sur les déplacements du
    hérisson européen (Erinaceus europaeus L.), Terre et Vie, 32:
    73-82.
Boissin, J., Daniel, J. Y., Ixart, G., Canivenc, R., 1975a, Etude
    préliminaire des glucocorticoïdes chez quelques mammifères
    sauvages, J. Physiol. (Paris) 71:126 A-127 A.
Boissin, J., Nouguier, J., Assenmacher, I., 1975b, Free ruhning

entrained and resynchronized rhythms of plasma corticosterone
and locomotor activity in quails, Int. J. Chronobiol. 3:89-125.
Boissin-Agasse, L., Maurel, D., Boissin, J., 1981, Seasonal variations in thyroxine and testosterone levels in relation to the
moult in the adult male mink (*Mustela vison* Peale and Beauvois),
Can. J. Zool., 59:1062-1066.
Boulouard, R., 1967, Activité corticosurrénalienne chez le Lérot
(*Eliomys quercinus*) durant la période de préparation à l'hiber-
nation, J. Physiol., (Paris) 59:221.
Dussart, G., Richoux, J. P., 1973, Régulation des hormones gonado-
tropes sur les activités monoamines oxydasiques diencéphaliques
et sur les glandes génitales durant l'hibernation, Ann. Endocr.
(Paris) 34:115-132.
Gabe, M., Agid, R., Martoja, A., Saint-Girons, M. C., Saint-Girons,
M., 1963, Données histophysiologiques et biochimiques sur l'
hibernation et le cycle annuel chez *Elyomis quercinus*, Arch.
Biol. (Liège) 75:1-87.
Herter, K., 1933, Gefangenschaftsblobachtungen an europaischen
Igeln II, Zeit. F. Säugetierkunde, 8:195-218.
Hoffman, R. A., Zarrow,M. X., A comparison of seasonal changes
and the effect of cold on the thyroid gland of the male rat
and ground squirrel (*Citellus tridecemlineatus*), Acta endocr.,
27:77-84.
Hudson, J. W., 1981, Role of the endocrine glands in hibernation
with special reference to the thyroid gland, in:"Survival in
the cold. Hibernation and other adaptations," X.J. Musacchia
and L. Jansky, eds., Elsevier/North Holland, Amsterdam.
Hudson, J. W., Wang, L. C., 1979, Hibernation: Endocrinologic
aspects, Ann. Rev. Physiol., 41:287-303.
Hulbert, A. J., Hudson, J. W., 1976, Thyroid function in a hiber-
nator *Spermophillus tridecemlineatus*, Am. J. Physiol., 230:
1211-1216.
Johansson, B. W., Senturia, J. B., 1972, Seasonal variations in the
physiology and biochemistry of the european hedgehog (*Erinaceus
europaeus*) including comparisons with non-hibernators, guinea-
pig and man, Acta Physiol. Scand.supp 380:1-159.
Kristoffersson, R., Soivio, A., 1967, A comparative long-term study
of hibernation in Finnish and German hedgehogs in a constant
ambient temperature, Ann. Acad. Sci. Fenn. (Ser. A IV) 122:1-23.
Lachiver, F., Olivereau, M., Kayser, C., 1957, L'activité de la
thyroïde chez un hibernant, le Lérot (*Eliomys quercinus* L.)
en hiver et au printemps, C.R. Soc. Biol., 151:653-656.
Maurel, D., Boissin, J., 1981, Plasma thyroxine and testosterone
levels in the red fox (*Vulpes vulpes* L.) during the annual
cycle, Gen. Comp. Endocr., 43:402-404.
Maurel, D., Joffre, J., Boissin, J., 1977, Cycle annuel de la testos-
téronémie et de la thyroxinémie chez la Blaireau européen
(*Meles meles* L.), C.R. Acad. Sci. (Paris), Ser. D 284:1577-1580.
Morris, P., 1973, Winter nests of the hedgehog (*Erinaceus europaeus*
L.), Oecologia, 11:299-313.

Musacchia, X. J., Jansky, L., eds., 1981, Survival in the cold. Hibernation and other adaptations, Elsevier/North Holland, Amsterdam.

Pengelley, E. T., Asmundson, S. J., 1974, Circannual rhythmicity in hibernating mammals, in:"Circannual clocks–Annual biological rhythms," E.T. Pengelley, ed., Academic Press

Petrović, A., Kayser, C., 1958, Variations saisonnières du seuil réactionnel de la thyroïde à la thyreostimuline chez le Hamster (*Cricetus cricetus*), J. Physiol. (Paris) 50:446-450.

Saboureau, M., 1979, Cycle annuel du fonctionnement testiculaire du Hérisson (*Erinaceous europaeus* L.). Sa régulation par les facteurs externes et internes. Thèse Doctorat ès sciences, Université Tours, 198 p.

Saboureau, M., Dutourné, B. 1981, The reproductive cycle in the male hedgehog (*Erinaceous europaeus* L.): a study of endocrine and exocrine testicular functions, Reprod. Nutr. Develop., 21: 109-216.

Saboureau, M., Laurent, G., Boissin, J., 1979, Daily and seasonal rhythms of locomotor activity and adrenal function in male hedgehog (*Erinaceus europaeus* L.), J. Interdiscipl. Cycle Res., 10:245-266.

Saboureau, M., Bobet, J. P., Boissin, J., 1980, Activité cyclique de la fonction corticosurrénalienne et variations saisonnières du métabolisme périphérique du cortisol chez un mammifère hibernant, le Hérisson (*Erinaceus europaeus* L.), J. Physiol. (Paris) 76:617-629.

Saure, L., 1969, Histological studies on the sexual cycle of the male hedgehog (*Erinaceus europaeus* L.), Aquilo, 9:1-43.

Suomalainen, P., Nyholm, P., 1956, Neurosecretion in hibernating hedgehog, in:"Zool. papers in honour of Bertel Hanstroms, 65th birthday.",Lund., pp. 265-277.

Szafarczyk, A., Boissin, J., Assenmacher, I., 1971, Effets du niveau d'éclairement sur le rythme circadien de la corticostéronémie chez la ratte, C.R. Acad. Sci., Paris, 273:2583-2586.

Vigouroux, E., 1972, Radiodosage ultrasensible de la thyroxine par la méthode de compétition, C.R. Acad. Sci, Paris, 275:579-582.

Walhovd, H., 1973, Observationer of pindsvin vinteren 1972-73, Flora og Fauna, 79:66-68.

Walhovd, H., 1975, Winter activity of danish hedgehogs in 1973-74, with information on the size of the animals observed and location of the recordings, Natura Jutlandica, 18:53-61.

Wimsatt, W. A., 1969, Some interrelations of reproduction and hibernation in mammals, in: "Dormancy and survival," A. H. W. Woolhouse, ed., Univeristy Press, Cambridge.

COMPARATIVE MECHANISMS OF PHYSIOLOGICAL, METABOLICAL AND ECO-ETHO-
LOGICAL ADAPTATION TO THE WINTER SEASON IN TWO WILD EUROPEAN
MAMMALS: THE EUROPEAN BADGER (*MELES MELES* L.) AND THE RED FOX
(*VULPES VULPES* L.)

D. Maurel and J. Boissin

Centre d'Etudes Biologiques des Animaux Sauvages
C.N.R.S.
Villiers-en-Bois, 79360 Beauvoir-sur-Niort, France

INTRODUCTION

In wild European mammals, the winter season, even in Southern
temperate regions, is a critical period for survival. In addition
to unfavourable climatic conditions (low temperatures, frost and
snow), natural sources of food are scarce and sometimes even disappear completely. Thus the animal has to apply a strategy of
adaptation that can be described as an etho-physiological strategy
as it involves both physiological and behavioral mechanisms of
adaptation. These mechanisms are part of the animal genetic
programme and enable it to adapt to variations in the surrounding
environment, thus ensuring survival. For one category of animals,
survival is assured through hibernation which means that the
animal is disconnected from its environment (Wang and Hudson, 1978)
whereas for others, survival is assured by the seasonal adjustment
of vital functions.

In our work, we considered two essential physiological
functions: the testicular function and the thyroid function in
two mammals typical of the woodland ecosystem, the European badger
(*Meles meles* L.) and the red fox (*Vulpes vulpes* L.). Reproduction
is directly involved in the application of demographic strategy
and thus in the survival of the species in as far as there is
strict synchronisation with seasonal rhythm; the thyroid function
is directly involved in the regulation of the metabolic processes
especially in the mechanism of thermoregulation either chemical
(production of energy) or physical (moulting and pelage) and in
the seasonal constitution and utilization of organic reserves;
when these mechanisms function adequately the survival of the
individual animal is assured.

Finally there also exists a specific type of adaptation concerning behavior which assures the survival of the individual animal and of the species through seasonal adjustments in activity rhythm, movements and habits.

MATERIAL AND METHODS

Study Area

This study was carried out at the CEBAS (Centre d'Etudes biologiques des Animaux Sauvages) in the Chizé Forest located in midwestern France (lat. 46° 07' N, Long. 0° 25' W); 2500 of the 5000 hectares total forest area are enclosed with three meter high fencing. The forest is primarily composed of beech and oak with a few areas of planted pine. The oceanic climate is temperate, the winters being relatively mild.

Animals

The animals, adult males, were captured in the forest near the laboratory. For the studies relating to free living animals the foxes and badgers after capture were equipped with transmitter-collars and then released in the forest . Their movements were automatically followed, day and night, their position recorded every four minutes by means of an automatic radio-tracking system set up at the CEBAS (Déat et al., 1980).

For the studies on activity rhythm and endocrine functions, the animals were weighed on arrival at the laboratory, examined and placed in individual pens in the forest so that they were exposed to natural variations in environmental factors; they were fed daily and water was provided *ad libitum*. Once or twice a month the animals were weighed and a blood sample was taken from the radial vein after slight anaesthesia to determine the endocrine parameters.

Study Techniques

Testicular activity was determined by testicular weight and plasma testosterone levels, measured by radioimmunoassay (Maurel et al., 1981). The metabolic parameters of testosterone were determined by an injection of ^{3}H-testosterone (Maurel and Boissin, 1982). Thyroid activity was determined, on the one hand, by determining plasma thyroxine using the method of radiocompetition described by Vigouroux (1972) and validated for the Badger and the Fox (Maurel and Boisin, 1979, 1981) and, on the other hand by the study of thyroid glandular uptake *in vivo* after injection of radioactive Iodine (^{131}I) and the study of the biological half-life (t 1/2) of thyroxine after injection or radioactive thyroxine

(^{131}I-L-T$_4$) (Maurel and Boissin, 1979, 1981). Activity rhythm and movements were determined using the fixed automatic radio-tracking system in the Chizé forest. The data were then fed into a Digital-Declab computer PDP 11-40 and analysed by means of programmes specially designed for these species (Déat et al., 1980). For the badgers kept in pens, the analysis of activity rhythm was done in a similar way; a system of infrared actographic apparatus was used (Maurel, 1981). These results concerning behavior were completed by a field study on burrows and frequented areas.

All results are expressed in the form of mean values shown with their standard error. Variance analysis (test F) was used for the comparison of the mean values.

RESULTS

Testicular Function

In the badger: the analysis of variations in testicular weight and the study of seasonal fluctuations in plasma testosterone show that winter is the main breeding period (Table 1 and Fig. 1a). During the winter months, from January to March, testicular weights are at their highest and, in addition, testosterone levels are at a maximum. Testicular activity continues until summer and high levels of testosterone and high testicular weights can still be observed in July. It is only in fact from mid-summer on that all the parameters relating to the testicular function are depressed. This general pattern in the sexual cycle is repeated year after year (Fig. 1a).

Table 1. Seasonal variations in testicular weights in the badger and in the fox. The weight of the testis is given as a seasonal mean value ± standard error.
The figure in brackets (n) indicates the number of animals examined (two tests for each animal).

Season	Testicular weight (in grammes)	
	Badger	Fox
Spring	6.5 ± 0.2 (12)	1.7 ± 0.4 (17)
Summer	5.9 ± 0.3 (6)	1.3 ± 0.1 (11)
Autumn	3.9 ± 0.3 (5)	2.3 ± 0.4 (15)
Winter	7.7 ± 0.8 (18)	5.2 ± 0.3 (21)

BADGER

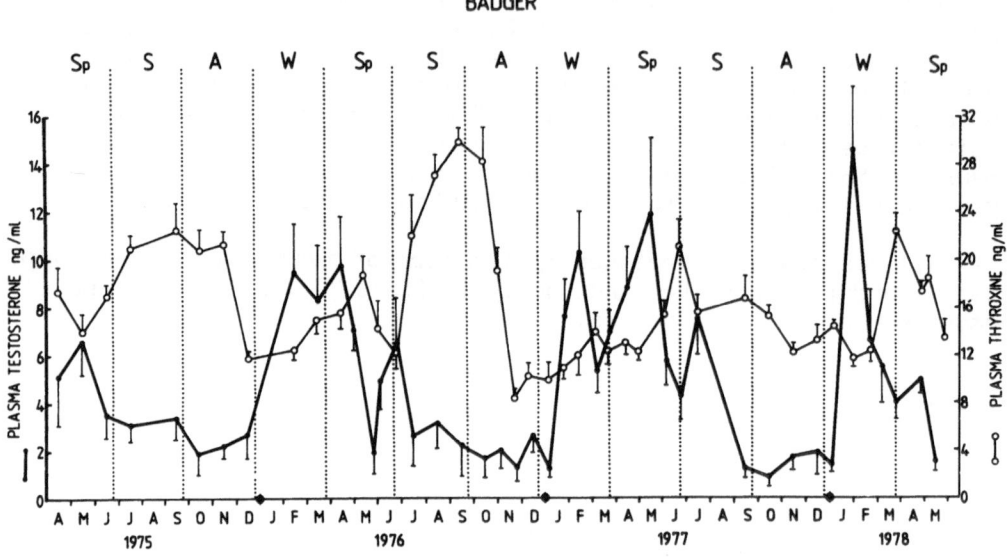

FOX

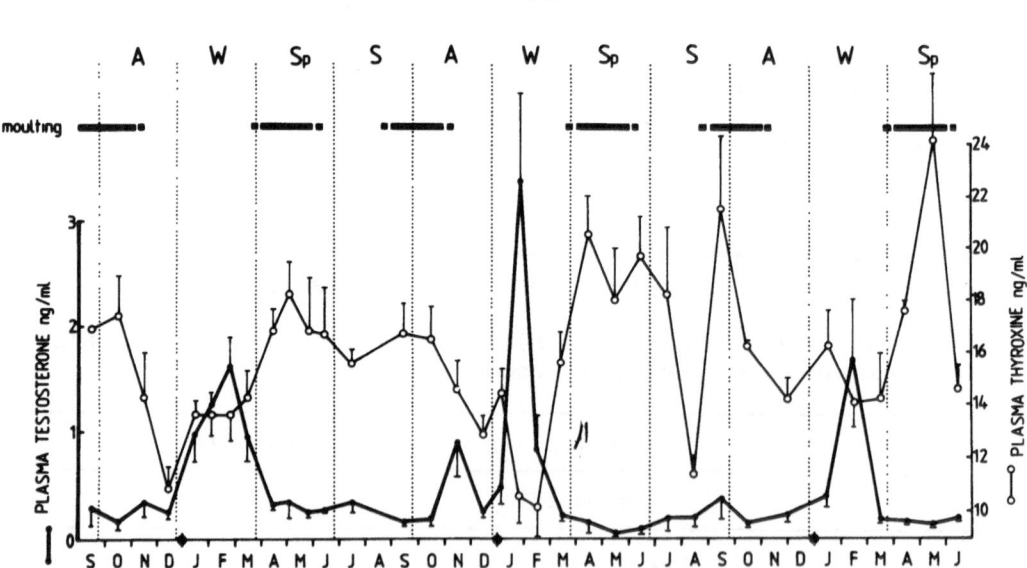

Fig. 1. Annual cycles of plasma testosterone and thyroxine in the
badger and in the fox over a period of several years in
relation to the moulting cycle. The coulting period in the
badger is similar to the fox but the phenomenon is less
intense.

In the fox: the seasonal aspect of the cycle is even more pronounced than in the badger. Thus, apart from the winter period, characterized by significantly high testicular weights and testosterone levels (Table 1 and Fig. 1b) there is no sexual activity, the phase of testicular endocrine and exocrine regression being equivalent to a seasonal castration from April to October.

Reproduction Cycle

In the badger: the period of pregnancy is exceptionally long, lasting one year due to the phenomenon of delayed ovo-implantation. The "post-partum" oestrus takes place from January to March (Canivenc, 1960). Thus the females are fecundated just after the birth of the young but the egg remains in the state of free blastocyst and implantation occurs only 10 months later, in November-December; so the actual gestation only lasts two months. The length of the reproductive season in the male, characterized by active spermatogenesis and still high levels of testosterone in July can be explained by another particularity of the female sexual cycle: subadult females, immature during the main winter rutting season in the male, present their first oestrus only from the fifteenth month on, that is to say from April onward (Ahnlund, 1980) and at the CEBAS we were able to obtain fertile sexual mating with nulliparous females from May to July.

It must be noted that the main mating period is extremely short im time as, for the adult female, it takes place in the few days following the parturition. The complete reproductive cycle is shown in Fig. 2.

In the fox: the annual reproductive cycle is clearly defined in time for both sexes. The oestrus of the female occurs exclusively in January-February (Rowlands and Parkes, 1935), thus in winter during the rutting season in the male. Outside the winter period, genital activity is minimal for both males and females (strict anoestrus).

Thyroid Activity

In the badger: if we consider all parameters, glandular as well as plasmatic, the winter season can be defined as a period of thyroid hypoactivity (Fig. 3a). Thyroid uptake measured 24 hours after injection of radioactive Iodine remains low from the end of autumn to the end of winter (winter mean value: $23\pm3\%$ vs summer-autumn mean value: $49\pm4\%$). Similarly, hormonal secretion, expressed by the factor K_4 is very low (winter mean: $1.8\pm0.3\%$ vs autumn mean: $10.1\pm2.4\%$). Finally, plasma levels of thyroxine are also at their lowest in winter, about 12 ng/ml, whereas the highest levels in summer and in early autumn can attain 25 to 30 ng/ml. Thus the seasonal thyroid function is characterized in the badger by a

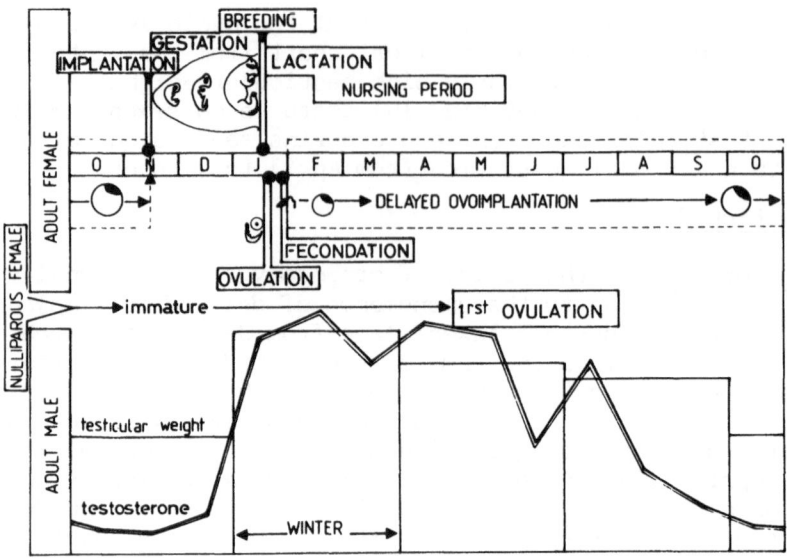

Fig. 2. Scheme of the cycle of reproduction in the European badger.

high level of activity at the end of summer and beginning of autumn, then a sharp decrease at the beginning of winter, followed by a winter period of hypoactivity; a brief increase takes place in spring and after a pre-summer depression, there is a summer recrudescence (Fig. 1a).

In the fox: the seasonal variations of the different parameters of thyroid activity are not so clearly marked as in the badger; we can no longer speak of winter thyroid hypoactivity but rather thyroid hyperactivity from spring to autumn (Fig. 3b). Thyroid uptake is constant at 35-40% and so a characteristic cycle can not be defined using this parameter alone. If we examine the factor of hormonal excretion (K_4) and the thyroxinemia, winter is a period of transition at the beginning of which a decrease in both parameters can be observed. However, starting from January, there is an increase in activity rising to a spring maximum which is regularly found each year (Fig. 1b).

Body Weight Cycle

In the badger: there exists a clearly defined seasonal variation in body weight (Fig. 4). From April to July, body weight is at its minimum ($\bar{m} = 9.9 \pm 0.1$ kg). From the end of summer a fattening can be observed and the maximum weight is reached at the beginning of winter (beginning of January: 12.5 ± 0.7 kg). Winter brings on loss of weight and body weights are at their lowest at the beginning of spring.

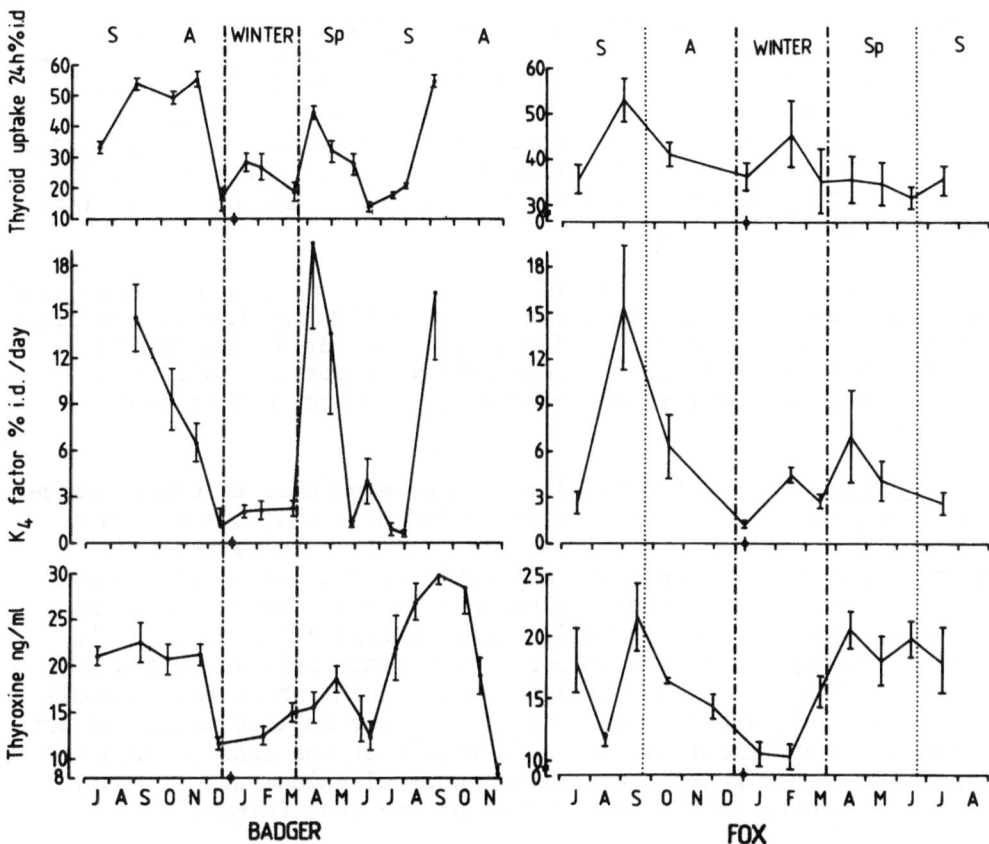

Fig. 3. Seasonal thyroid activity in the badger and in the fox.
1. Maximum thyroid uptake (24 h after injection of tracer) expressed in % uptake of injected dose (% i.d.).
2. Coefficient of glandular hormonal excretion (factor K_4) expressed in % of the injected dose per 24 hours (% i.d./ 24 h).
3. Plasma thyroxine levels (ng/ml).

In the fox: one can hardly detect a seasonal cycle of body weights; by regrouping the complete set of mean values measured over a period of five years, a significant increase in January and February can be observed (Fig. 4). However, this increase is very slight and the difference between the maximum mean (6.2±0.1 kg) and the annual mean value (5.8±0.1 kg) is no higher than 400 g.

Metabolic Activity

In the badger: peripheral utilization of thyroxine and testos-

terone varies according to season. A considerable decrease in
hormonal metabolism can be observed during winter. Thus, the biological half-life (t 1/2, hour) of thyroxine is 21.7$\pm$0.5 h in
winter whereas in spring the mean value of this parameter is
16.4$\pm$1.2 h; similarly the metabolic clearance rate of testosterone
(MCR, 1/24 h/kg) is significantly low in winter (70$\pm$5 1/24 h/kg)
in relation to autumn (112$\pm$13 1/24 h/kg) or to spring (117$\pm$6 1/24
h/kg) (Fig. 5).

Study of the metabolism of the fat cell shows that a high lipolysis is observed which is coupled with a slowing down of the lipogenesis in winter, period during which the animal uses the fats
accumulated in autumn when lipogenesis is important. All these
factors lead to a marked and characteristic body weight cycle
(Fig. 4).

In the fox: the few results that we were able to obtain concerning the peripheral use of thyroxine indicate that the seasonal
variations of the biological half-life of this hormone are not
significant and therefore are not worth discussing at the present
time. However, the metabolic clearance rate of testosterone increases sharply in winter (MCR: 81.2$\pm$1.2 1/24 h/kg) compared to
spring, summer and autumn values, which range between 30 and 50
1/24 h/kg (Fig. 5). The only interesting point about the metabolism
of the fat cell is the existence of a winter maximum value for both
lipogenesis and lipolysis but as a result of the less important
lipolytic than lipogenic activity, ponderal values are slightly
higher in winter than at other time of the year (Fig. 4).

General Activity and Movements

Nycthemeral and seasonal rhythm of locomotor activity. In the
badger and the fox, the synchronization of the nycthemeral rhythm
of generalized locomotor activity with sunset and sunrise has been
clearly shown, whether the animals are kept in outdoor pens or
living freely in their natural environment. The activity phase is
exclusively nocturnal in the badger (Fig. 6).

In the fox, activity begins earlier (twilight period) and ends
later (morning period); the rhythm, biphasic in the badger, is polyphasic with phases of locomotor activity during day time (Maurel,
1980). For the fox, a significant seasonal modification in activity
rhythm can not be shown, whereas for the badger the different parameters of general locomotor activity reveal a seasonal pattern
characterized by a sharp decrease in winter (Fig. 7). At the beginning of winter, activity in the badger is drastically reduced and
the animal may remain underground for as long as 10 days without
emerging for food or water.

Specific activity and movements. There exists in the badger a
specific activity (burrowing) which is typically seasonal as it is

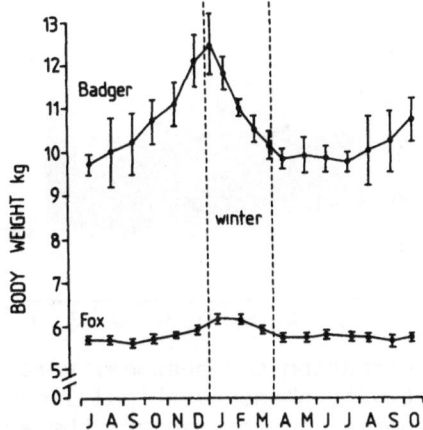

Fig. 4. Seasonal variations in body weight in the badger and the fox; once or twice monthly mean values (± s.e.m.) calculated over a period of five years.

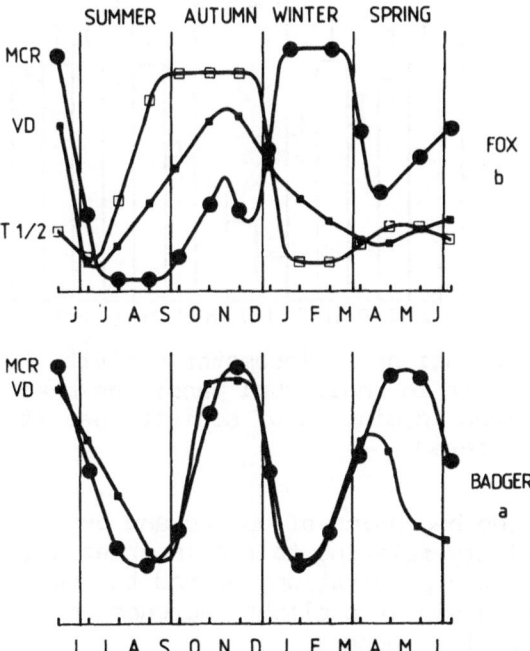

Fig. 5. Seasonal variations in various parameters of the peripheral metabolism of testosterone in the badger and in the fox: biological half-life (t 1/2), metabolic clearance rate (MCR) and volume of distribution (VD) of the male hormone.

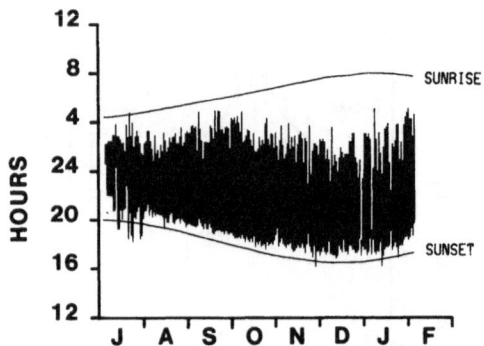

Fig. 6. Nycthemeral situation and seasonal variations in the daily
        locomotor activity phase (vertical lines) in relation to
        sunset and sunrise in the badger; the example is that of
        an animal kept in an individual pen.

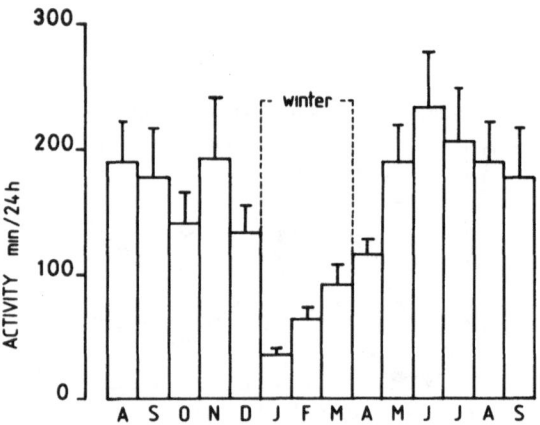

Fig. 7. Seasonal variations in locomotor activity for a group of
        5 badgers kept in individual pens; the mean values (± s.e.
        m.) are given in minutes of activity per 24 hours for each
        month considered.

observed only at the beginning of spring and beginning of autumn.
This specific activity relating to the interior organization of
their burrow marks the preparation for and the end of wintering in
this species; during winter a slight recrudescence is noted just
before the birth of the young.

     No activity of this type is observed in the fox which uses
natural burrows or shelters with little or no re-arranging.

Movements and space use. Although both species cover a lot of space, the fox (from 500 to 2000 ha, depending on individual animals) covers more space than the badger (200 ha on average). The badger has only one burrow situated usually in a central position in its area whereas the fox has one or several main burrows situated on the border limits of the home range and numerous secondary shelters or resting places scattered throughout the area frequented by the animal. The analysis of individual movements enable us to show that the fox possesses a network of preferential passages, situated either inside its home range or in certain border areas marking the boundaries of its territory. The intense activity of passage and marking deposits of urine and feces in the winter rutting and reproductive period correspond to a territorial activity on the boundaries (Maurel, 1982). The establishment of hierarchical relationships concerning eventual contacts on boundaries shows that there exists a strict territoriality in this species in which males exclude each other.

In the badger, the appearance in the winter period of far-ranging movements leading the animal to the limit of its area (Fig. 8) shows the existence of a territorial activity with fecal dropping in this species also; however, it is less exclusive than in the fox. Moreover, this territorial activity is characteristic of all the clan which can include several male and female adults. It must be stressed that all the parameters of general activity are depressed in winter and the number of movements greatly reduced, the few movements observed are far ranging and certainly related to the reproductive period.

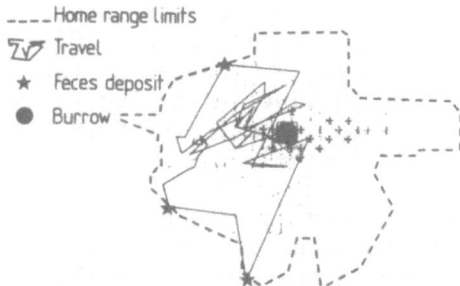

Fig. 8. Example of movements during 1 night (in the winter rutting period) in the badger.

DISCUSSION

The manifestation of reproductive activity and the considerable
energetic needs, which the accomplishment of this function
necessitates, whether on the level of the search for sexual partner
or on the level of the establishment of the numerous organic
correlations involved in the activity of the genital glands, are
in competition with other energetic needs relating to the survival of the individual animal: alimentary needs and protection
in relation to thermal variations of the environment.

Thus during the cold winter season, the badger can remain
underground for several days on end, reducing considerably its
locomotor activity; the increase in the length of time spent
underground in winter means that the animal is only exposed
incidentally to the action of low external temperatures, the
temperature of the burrow or that of the chambers lined with dried
grass being relatively constant and usually about 10°C in winter.
In the badger, the period of reduced activity starts in autumn
and continues through to spring. During this wintering, testicular
activity is at its maximum and for the female, the parturition and
"post partum" oestrus take place as well as successful mating. It
is important to distinguish wintering from hibernation; during
the latter, much more so than in wintering, the metabolic processes
are considerable reduced, and there is an important decrease in
central temperature which does not occur in wintering. This period
of reduced general activity also characterized by irregular
feeding, can not occur unless there is metabolic preparation in
the form of accumulation of organic reserves. This accumulation
is to be found in the badger as well as in hibernators but fattening only begins in autumn for the badger whereas the process begins
as early as June for the hedgehog (Saboureau, 1979). This phase
of autumn fattening takes place owing to a decrease in spontaneous
or "induced" lipolysis (Chraibi et al., 1982). During wintering
all the parameters characterizing thyroid activity are at their
lowest level and even if testosterone is produced in quantity
(Maurel and Boissin, 1982) the metabolism of the male hormone is
reduced: the metabolic clearance rate decreased and protein-
binding capacity increased (Maurel et al., 1980). The only intense
metabolic activity in winter is observed in fat cells with a
maximal lipolysis. Thus, in winter, all functions facilitate the
reduction in general metabolic activity, only the reproductive
function is at its highest level.

In the fox, however, the absence of a wintering period (no
winter hypothyroidism, no reduction in locomotor activity) and the
absence of modifications in the lipid metabolism (no seasonal
fattening, few seasonal variations in lipolysis) implies a constant search for food and thus an important locomotor activity;
in the same way, the polyphasic nycthemeral rhythm of general

activity, the absence of fixation to a well-structured burrow and the fact that the earths or shelters only offer the animal slight protection necessitate the presence of a winter fur which will prevent thermal exchanges and a summer fur which will on the contrary facilitate thermal exchanges. It must be noted that the two moulting periods are not situated during the reproductive season, but during the phase of maximum thyroid activity. It would seem that the animal can not use its energy for both protection and reproduction at the same time. The existence of seasonal variations of physiological functions, involved in all the processes of adaptation, perimits a metabolic anticipation which ensures the survival of the individual animal and the survival of the species.

From the behavioral point of view, social grouping and fixation (or non-fixation) to a burrow also proceed from a phenomenon relating to adaptation. Indeed, as the badger, lives in a set of family clan (Kruuk, 1978) with the same mate for several years and in a fixed, well-structured burrow, the search for a sexual partner and territorial acquisition are avoided and the reproductive cycle can take place in favourable conditions even although the individual animal is metabolically at a disadvantage because of a compulsary phase of wintering and, in particular, the female can be fecundated only just after parturition, when its locomotor activity is greatly reduced. The territory marking-activity is done by all the badgers forming the same clan and so necessitates a small amount of individual energy.

The fox, however, must acquire and maintain its territory and, at the same time, continue its search for food; so a thick winter fur prevents caloric loss by limiting thermal exchanges. The manner of social grouping which may differ, in the fox, according to the biotopes (bibl. in Maurel, 1981) is generally based on an independent, transitory couple which can accept other non-reproductive females (MacDonald, 1980). This couple cares for the young and breaks up in the autumn of the same year, when the young are able to fend for themselves.

Thus, on the basis of these observations (social life groups, general polyphasic locomotor activity and in particular the absence of significant seasonal variations in this activity for the fox, wintering period in the badger...) we can say that the fox is characterized by a greater plasticity and that it shows a greater adaptability to environmental factors than the badger which, being perfectly adapted to its environment, seems to be less able to cope with eventual changes or disruptions in its biotope. However, if we accept the view that the most significant sign of adaptability in a species is the duration of the reproduction period, we have to admit that the fox with its limited reproductive period profits from its capacity to adapt only on the individual level.

REFERENCES

Ahnlund, H., 1980, Sexual maturity and breeding season of the bad-
    ger, *Meles meles*, in Sweden, J. Zool., 190:77-95.
Canivenc, R., 1960, L' ovoimplantation différée des animaux sauvages,
    in:"Les fonctions de nidation utérine et leurs troubles," R.
    Canivenc, ed., Masson, Paris.
Chraibi, F., Desbal, B., Pejoan, C., Saboureau, M., Maurel, D., and
    Boissin, J., 1982, Variations saisonnières de la lipolyse des
    adipocytes de Renard, de Blaireau et de Hérisson. Relation
    avec les cycles annuels des activités testiculaire et thyro-
    ïdienne, J. Physiol.,Paris, 78:207-213.
Déat, A., Mauget, C., Mauget, R., Maurel, D., and Sempéré, A., 1980,
    The automatic continuous and fixed radiotracking system of
    the Chizé forest: theoretical and practical analysis, in:
    "Handbook on biotelemetry and radiotracking,"C.J. Amlaner and
    D.W. MacDonald, eds., Pergamon Press
Kruuk, H., 1978, Spatial organization and territorial behavior of
    the European badger, *Meles meles*, J. Zool.,London, 184:1-19.
MacDonald, D. W., ed., 1980, "Rabies and wildlife," Oxford Uni-
    versity Press, Oxford.
Maurel, D., 1980, Home range and activity rhythm of adult male foxes
    during the breeding season, in:"Handbook on biotelemetry and
    radiotracking," C.J. Amlaner and D.W. MacDonald, eds.,
    Pergamon Press
Maurel, D., 1981, Variations saisonnières des fonctions testiculaire
    et thyroïdienne en relation avec l'utilization de l'espace et
    du temps chez le Blaireau européen (*Meles meles*) et le Renard
    roux (*Vulpes vulpes*). Thèse de Doctorat ès-science, Université
    de Montpellier, France.
Maurel, D., 1982, Movements and space utilization in the fox (*Vulpes
    vulpes*) as studied by radiotracking. Proc. XV Cong. Int. Fauna
    Cineg. Silv., Caceres, Espagne.
Maurel, D., and Boissin, J., 1979, Seasonal variations of the
    thyroid activity in the adult male badger, *Meles meles*, Gen.
    Comp. Endocrinol., 38:207-214.
Maurel, D., and Boissin, J., 1981, Plasma thyroxine and testosterone
    levels in the Red fox, *Vulpes vulpes*, during the annual cycle,
    Gen. Comp. Endocrinol., 43:402-404.
Maurel, D., and Boissin, J., 1982, Métabolisme périphérique de la
    testostérone en relation avec le cycle annuel de la testos-
    térone et de la 5α-dihydrotestostérone plasmatique, chez le
    Blaireau européen (*Meles meles*) et le Renard roux (*Vulpes
    vulpes*), Can. J. Zool., 60:406-416.
Maurel, D., Laurent, A. M., and Boissin, J., 1981, Short-term
    variations of plasma testosterone concentrations in the
    European badger (*Meles meles*), J. Reprod. Fert., 61:53-58.
Maurel, D., Laurent, A. M., Daniel, J. Y., and Boissin, J., 1980,
    Etude de la capacité de liaison de la protéine plasmatique
    liant la testostérone chez deux Mammifères sauvages à activité

testiculaire cyclique, le Renard et le Blaireau, C.R. Acad. Sci. Paris, 291:693-696.

Rowlands, I. W., and Parkes, A. S., 1935, The reproduction processes of certain mammals. VIII reproduction in foxes (*Vulpes vulpes*). Proc. Zool. Soc. London.

Saboureau, M., 1979, Cycle annuel du fonctionnement testiculaire du Hérisson (*Erinaceus europaeus*). Sa régulation par les facteurs externes et internes. Thèse de Doctorat ès-science, Université de Tours, France.

Vigouroux, E., 1972, Radiodosage ultrasensible de la thyroxine par la méthode de compétition, C.R. Acad. Sci., 275:579-582.

Wang, C. H., and Hudson, J. W., eds.. 1978, "Strategies in cold: Natural torpidity and thermogenesis," Academic Press, New York.

AUTHOR INDEX

SYSTEMATIC INDEX

SUBJECT INDEX